Haruto

Akari

Sara

Yu

Study with your Friends!

How do we learn mathematics?

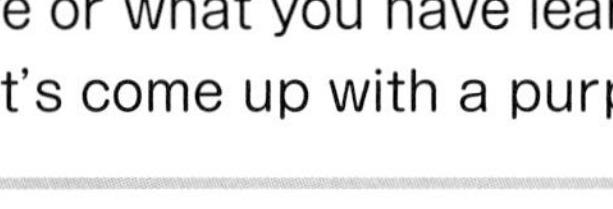

Based on the problem you find in your daily life or what you have learned, let's come up with a purpose.

1

The first problem of the lesson is written. On the left side, what you are going to learn from now on through the problem is written.

When you see the problem and think that you "want to think", "want to represent", "want to know", and "want to explore", that will be your "purpose" of your learning. You can find the purpose not only at the beginning of the lesson but in various timings and settings.

1

You can check your understanding and try more using what you have learned.

①

Let's try this problem first.

The starting point

What you have learned today

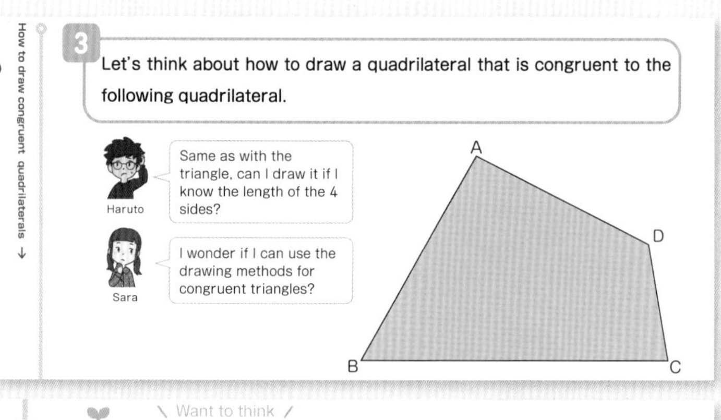

How to draw congruent quadrilaterals
3 Let's think about how to draw a quadrilateral that is congruent to the following quadrilateral.
Same as with the triangle, can I draw it if I know the length of the 4 sides?
Haruto
I wonder if I can use the drawing methods for congruent triangles?
Sara
A
D
B
C

\ Want to think /

? Purpose How can we draw a congruent quadrilateral?

1 Let's draw a quadrilateral that is congruent to the quadrilateral shown on the right.

Which sides and angles should we use?

2 Let's solve the following calculations in vertical form.

① 60×4.7　② 50×3.9　③ 7×1.6

④ 6×2.7　⑤ 24×3.3　⑥ 13×2.8

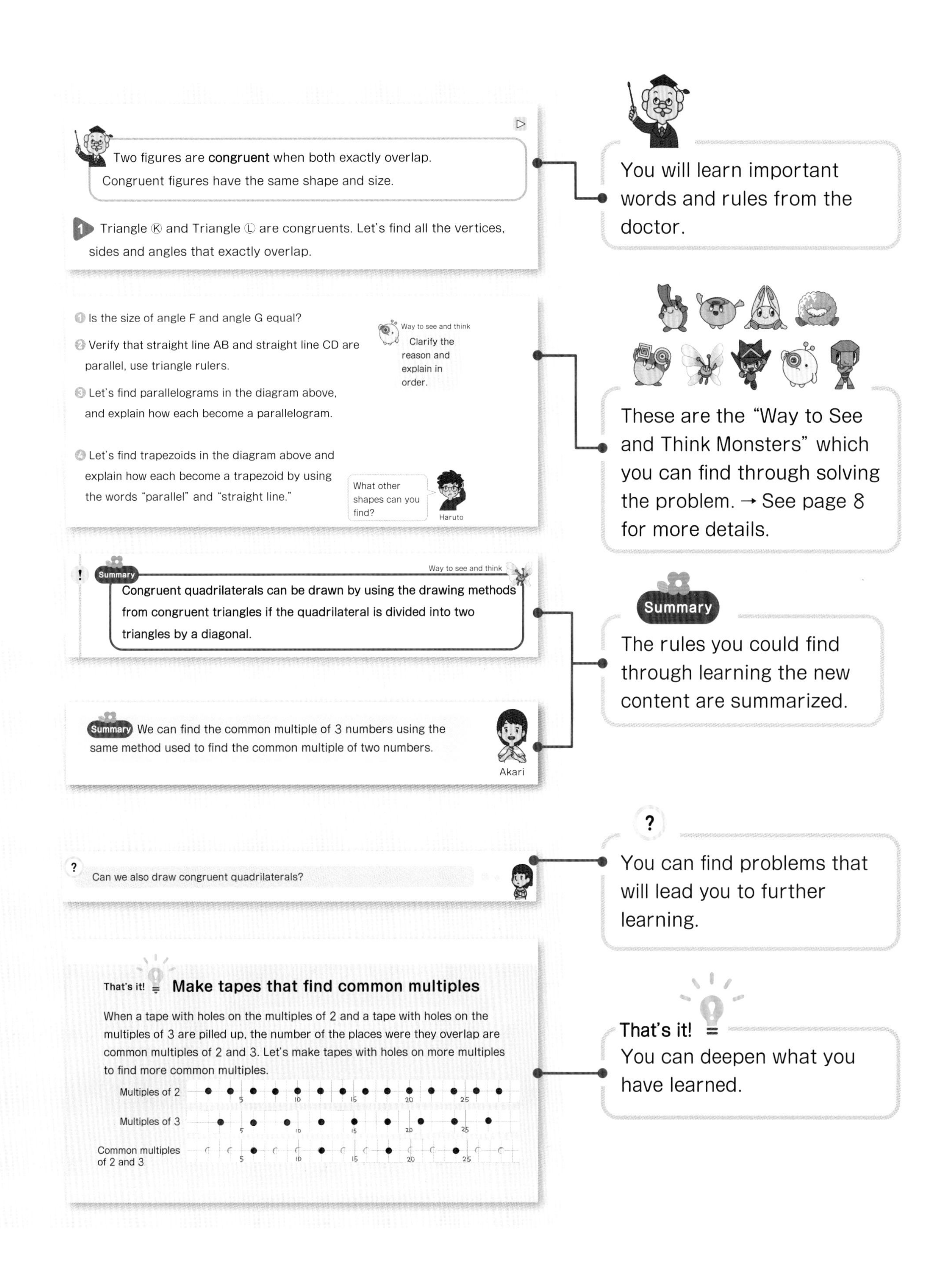
Two figures are congruent when both exactly overlap.
Congruent figures have the same shape and size.
1 Triangle Ⓚ and Triangle Ⓛ are congruents. Let's find all the vertices, sides and angles that exactly overlap.
You will learn important words and rules from the doctor.
① Is the size of angle F and angle G equal?
② Verify that straight line AB and straight line CD are parallel, use triangle rulers.
③ Let's find parallelograms in the diagram above, and explain how each become a parallelogram.
④ Let's find trapezoids in the diagram above and explain how each become a trapezoid by using the words "parallel" and "straight line."
Way to see and think
Clarify the reason and explain in order.
What other shapes can you find?
Haruto
These are the "Way to See and Think Monsters" which you can find through solving the problem. → See page 8 for more details.
Summary
Way to see and think
Congruent quadrilaterals can be drawn by using the drawing methods from congruent triangles if the quadrilateral is divided into two triangles by a diagonal.
Summary
The rules you could find through learning the new content are summarized.
Summary We can find the common multiple of 3 numbers using the same method used to find the common multiple of two numbers.
Akari
?
Can we also draw congruent quadrilaterals?
?
You can find problems that will lead you to further learning.
That's it!
Make tapes that find common multiples
When a tape with holes on the multiples of 2 and a tape with holes on the multiples of 3 are pilled up, the number of the places were they overlap are common multiples of 2 and 3. Let's make tapes with holes on more multiples to find more common multiples.
Multiples of 2
Multiples of 3
Common multiples of 2 and 3
5
10
15
20
25
That's it!
You can deepen what you have learned.

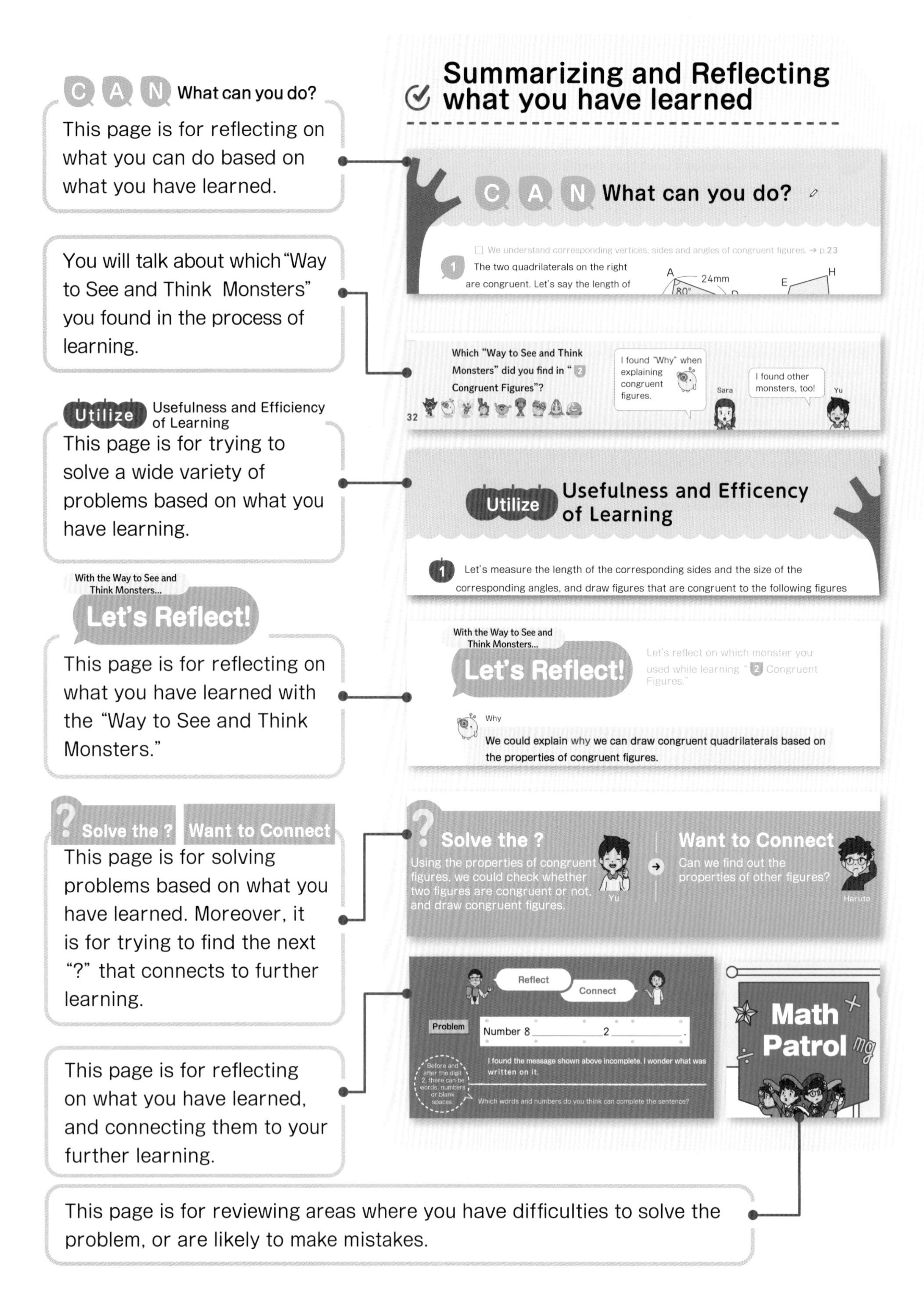

C A N What can you do?
This page is for reflecting on what you can do based on what you have learned.
You will talk about which "Way to See and Think Monsters" you found in the process of learning.
Utilize Usefulness and Efficiency of Learning
This page is for trying to solve a wide variety of problems based on what you have learning.
With the Way to See and Think Monsters...
Let's Reflect!
This page is for reflecting on what you have learned with the "Way to See and Think Monsters."
Solve the ?
Want to Connect
This page is for solving problems based on what you have learned. Moreover, it is for trying to find the next "?" that connects to further learning.
This page is for reflecting on what you have learned, and connecting them to your further learning.
This page is for reviewing areas where you have difficulties to solve the problem, or are likely to make mistakes.
Summarizing and Reflecting what you have learned
C A N What can you do?
We understand corresponding vertices, sides and angles of congruent figures. → p.23
1 The two quadrilaterals on the right are congruent. Let's say the length of
Which "Way to See and Think Monsters" did you find in " 2 Congruent Figures"?
I found "Why" when explaining congruent figures.
Sara
I found other monsters, too!
Yu
32
Utilize Usefulness and Efficency of Learning
1 Let's measure the length of the corresponding sides and the size of the corresponding angles, and draw figures that are congruent to the following figures
With the Way to See and Think Monsters...
Let's Reflect!
Let's reflect on which monster you used while learning " 2 Congruent Figures."
Why
We could explain why we can draw congruent quadrilaterals based on the properties of congruent figures.
Solve the ?
Using the properties of congruent figures, we could check whether two figures are congruent or not, and draw congruent figures.
Yu
Want to Connect
Can we find out the properties of other figures?
Haruto
Reflect
Connect
Problem
Number 8 2 .
I found the message shown above incomplete. I wonder what was written on it.
Before and after the digit 2, there can be words, numbers or blank spaces.
Which words and numbers do you think can complete the sentence?
Math Patrol

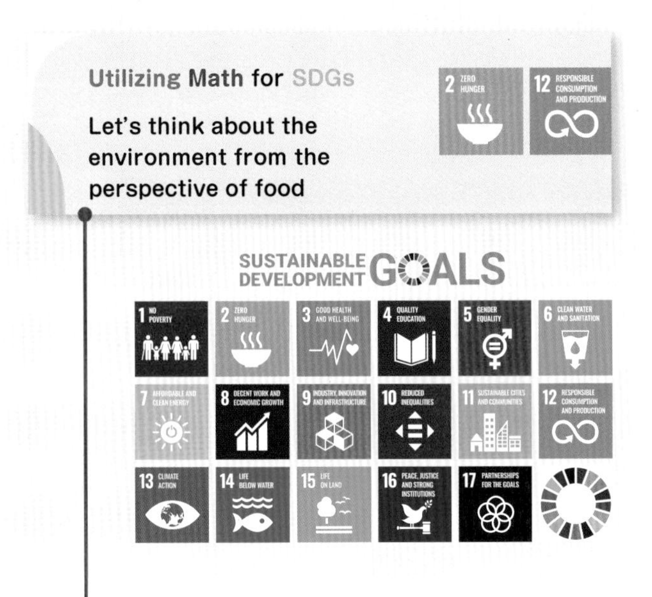

Utilizing Math for SDGs

The Sustainable Development Goals (SDGs) are a set of goals that we aim to achieve in order to create a world where we can live a life of safety and security. This page will help you think about what you can do for society and the world through math.

About the QR Code

Some of the pages include the QR code which is shown on the right.

- …You can learn how to draw a diagram and how to calculate by watching a movie.
- …You can learn by actually moving and operating the contents.
- …You can learn by reflecting on what you have learned previously in your previous grades.
- …You can utilize it to know the solution to the problems that you couldn't find out the answer, or to try various problems.
- …You can deepen your learning by actually looking at the materials including the website.

Dear Teachers and Parents

This textbook has been compiled in the hope that children will enjoy learning through acquiring mathematical knowledge and skills.
The unit pages are carefully written to ensure that students can understand the content they are expected to master at that grade level.
In addition, the "More Math" section at the end of the book is designed to ensure that each student has mastered the content of the main text, and is intended to be handled selectively according to the actual conditions and interests of each child.
We hope that this textbook will help children develop an interest in mathematics and become more motivated to learn.

The sections marked with this symbol deal with content that is not presented in the Courses of Study for that grade level, thus does not have to be studied uniformly by all children.

QR codes are used to connect to Internet content by launching a QR code-reading application on a smartphone or tablet and reading the code with a camera.The QR Code can be used to access content on the Internet.

https://r6.gakuto-plus.jp/s5a01

Note: This book is an English translation of a Japanese mathematics textbook. The only language used in the contents on the Internet is Japanese.

Infectious Disease Control
In this textbook, pictures of activities and illustrations of characters do not show children wearing masks, etc., in order to cultivate children's rich spirit of communicating and learning from each other. Please be careful to avoid infectious diseases when conducting classes.

Becoming a Writing Master

The notebook can be used effectively.

- To organize your own thoughts and ideas
- To summarize what you have learned in class
- To reflect on what you have learned previously

Let's all try to become notebook masters

Write today's date. → November 30th

Write the problem of the day that you must solve. → Problem

What is the area of the following figure in cm^2.

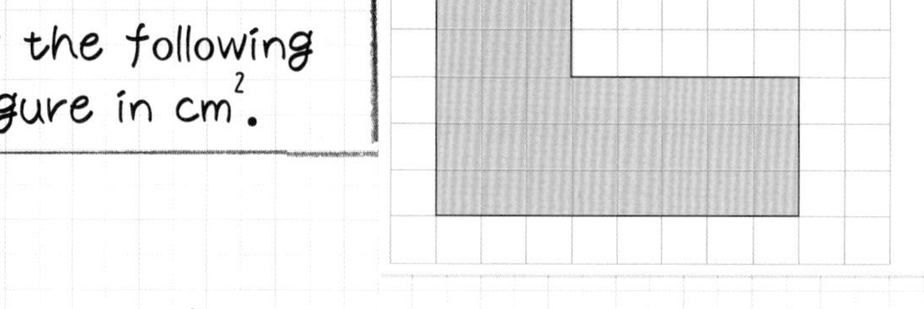

Let's write down what you thought while thinking about the solution of the problem as "purpose." → <Purpose>

Can we find out the area of a figure that is neither a rectangle nor a square?

Write your ideas or what you found about the problem. → O My idea

I divided the figure into two rectangles, and calculated.

$5 \times 3 = 15$

$3 \times 5 = 15$

$15 + 15 = 30$

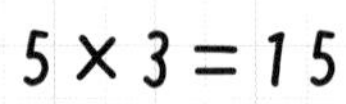

Answer: 30 ~~cm~~ cm^2

Tips for Writing 2

Where did I find this before?

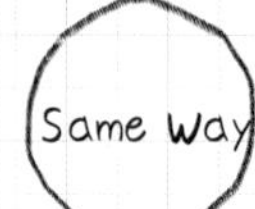

The calculation above can be summarized into one math equation.

$5 \times 3 + 3 \times 5 = 30$

Tips for Writing ①

When you made a mistake, don't erase it so that it will be easier to understand when you look back at your notebook later.

Tips for Writing ②

By finding the "Way to See and Think Monsters," it will connect you to what you have learned previously.

Tips for Writing ③

By writing down what you would like to try more, it will lead you to further learning.

Yu's idea
I assumed it to be a big rectangle and subtracted the dented part.

1 cm
1 cm

$5 \times 8 - 2 \times 5 = 40 - 10$
$= 30$
Answer: $30 cm^2$

← Write the classmate's ideas you consider good.

<Summary>

Even if it is not a rectangle or a square, by subtracting the dented part from the bigger rectangle, we can find out the area.

← Summarize what you have learned today.

<Reflection>
By changing the unknown figure to a figure we know how to find out the area, we could find out the area.

<What I want to do next>
I want to find out the area of various figures.

← Reflect on your class, and write down the following;
- What you learned.
- What you found out.
- What you can do now.
- What you don't know yet.

While learning mathematics...

Based on what I have learned previously...

Why does this happen?

There seems to be a rule.

You may be in situations like above. In such case, let's try to find the "Way to See and Think Monsters" on page 9. The monsters found there will help you solve the mathematics problems. By learning together with your friend and by finding more "Way to See and Think Monsters," you can enjoy learning and deepening mathematics.

What can we do at these situations?

I think I can use 2 different monsters at the same time...

→ You may find 2 or 3 monsters at the same time.

I came up with the way of thinking which I can't find on page 9.

→ There may be other monsters than the monsters on page 9. Let's find some new monsters by yourselves.

Now let's open to page 9 and reflect on the monsters you found in the 4th grade. They surely will help your mathematics learning in the 5th grade!

Representing ways of thinking in mathematics

Way to See and Think Monsters

Unit

If you set the unit...

Once you have decided one unit, you can represent how many using the unit.

Summarize

If you try to summarize...

It makes it easier to understand if you summarize the numbers or summarize in a table or a graph.

Other Way

If you represent in other ways...

If you represent in other Something depending on your purpose, it is easier to understand.

Align

If you try to align...

You can compare if you align the number place and align the unit.

Change

If you try to change the number or the figure...

If you try to change the problem a little, you can understand the problem better or find a new problem.

Divide

If you try to divide...

Decomposing numbers by place value and dividing figures makes it easier to think about problems.

Why

You wonder why?

Why does this happen? If you communicate the reasons in order, it will be easier to understand for others.

Rule

Is there a rule?

By examining, you can find rules and think using rules.

Same Way

Can you do it in a similar way?

If you find something the same or similar to what you have learned, you can understand.

Ways to think learned in the 4th grade

Number and Calculations

Figure

Align

You can compare and calculate by aligning the place value.

213610000 > 21363000

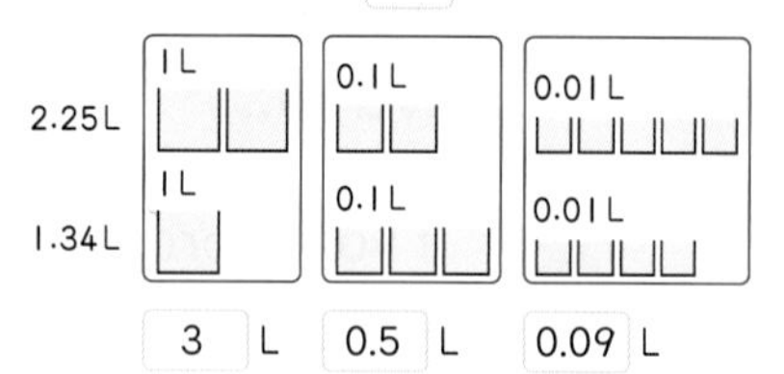

Unit

By setting a unit, you can represent the size of angles and areas with numbers.

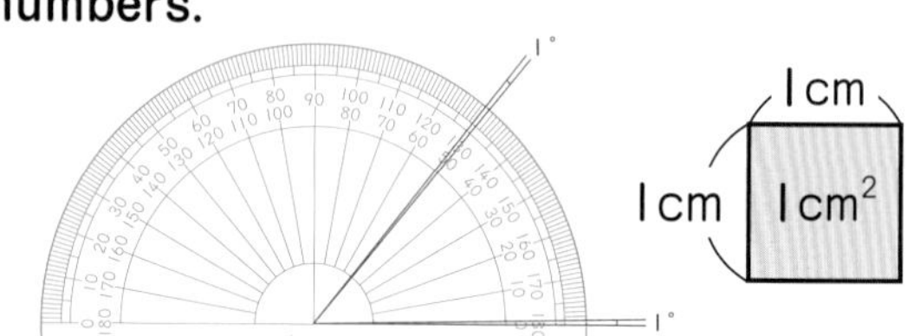

Rule

Finding the rules of division.

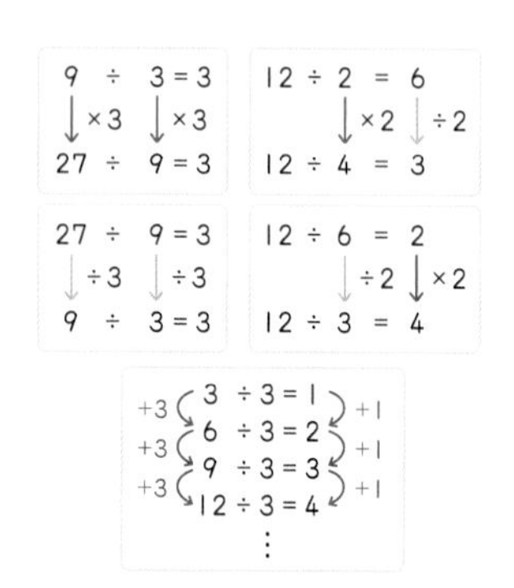

Divide

Decompose the dividend so that you can calculate division.

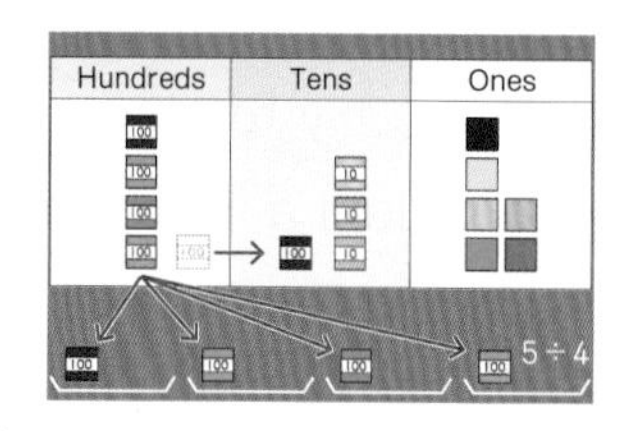

Divide

If we divide into squares, rectangles, etc., we can find out the area.

I count the number of 1-cm^2 squares.

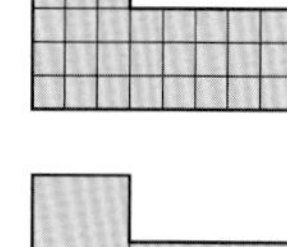

I calculate by separating the figure into 2 rectangles.

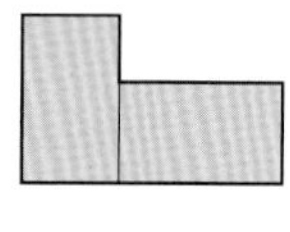

Unit

By setting 0.1 or $\frac{1}{8}$ as one unit, we can calculate in the same way as whole numbers.

× 3 =

× 3 =

12 set of 0.1　36 set of 0.1

$\frac{9}{8} - \frac{4}{8} = \frac{5}{8}$

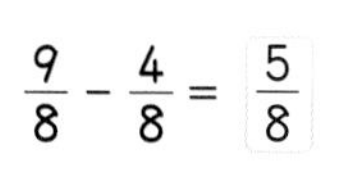

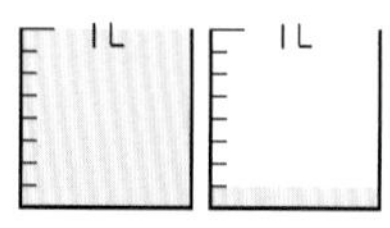

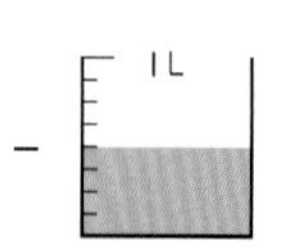

Summarize

We can classify quadrilaterals based on the relationship between the sides and how they intersect.

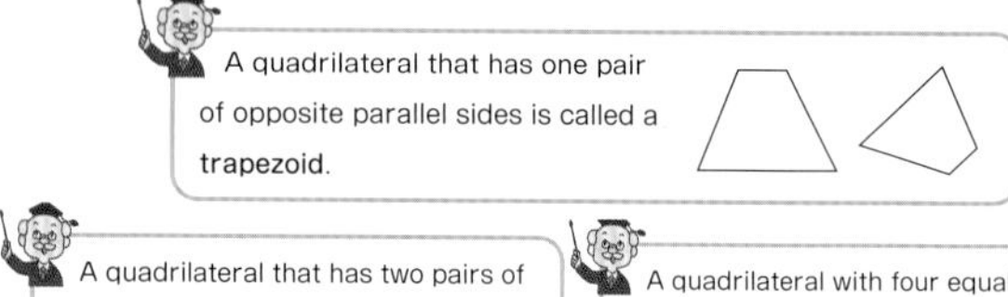

A quadrilateral that has one pair of opposite parallel sides is called a trapezoid.

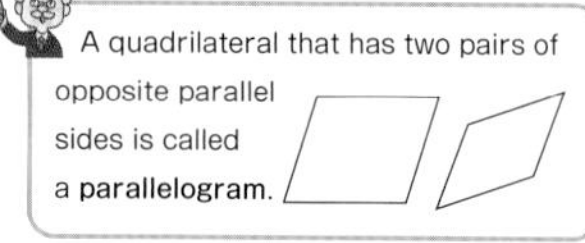

A quadrilateral that has two pairs of opposite parallel sides is called a parallelogram.

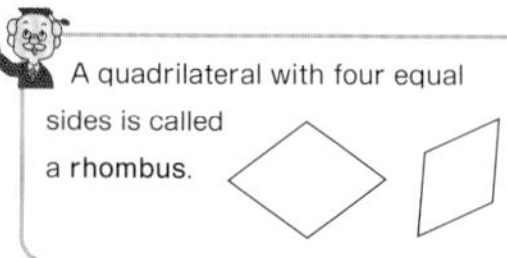

A quadrilateral with four equal sides is called a rhombus.

Data

Change and Relationship

Other way

Consider whether to see the two quantities as a difference or a multiple.

Rule

Find out the rule by representing quantities changing together in a table.

+1 +1 +1

Number of steps	1	2	3	4
Length(cm)	4	8	12	16

+4 +4 +4

×2 ×2

Number of steps	1	2	3	4
Length(cm)	4	8	12	16

×2 ×2

×4 ×4 ×4 ×4

Number of steps	1	2	3	4
Length(cm)	4	8	12	16

Other Way

Make a graph that is easy to understand according to your purpose.

Average temperature per month in Niigata City and Naha City

Month	1	2	3	4	5	6	7	8	9	10	11	12
Niigata City (℃)	5	5	8	10	17	22	24	28	24	16	11	5
Naha City (℃)	19	19	20	20	25	28	29	29	28	26	23	19

Unit

Change the size of one scale according to the purpose.

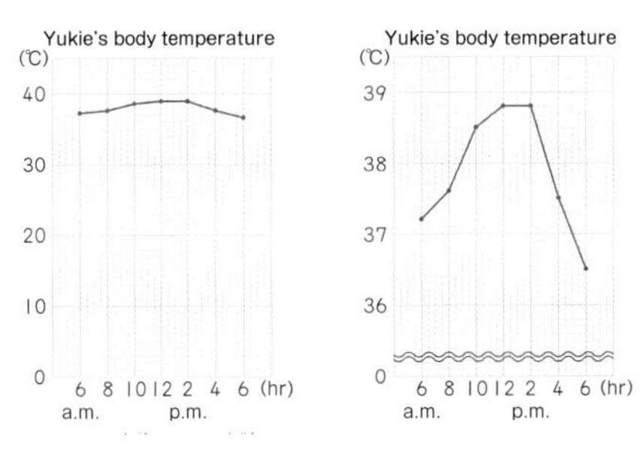

Summarize

Summarize and arrange two perspectives into one table.

Places and kind of injuries (children)

Place \ Kind	Cut	Bruise	Scratch	Sprained finger	Sprain	Total
Playground	0	3	4	1	1	9
Corridor	1	3	0	0	0	4
Classroom	3	0	3	0	0	6
Gymnasium	0	1	3	1	2	7
Stairs	0	1	1	0	0	2
Total	4	8	11	2	3	28

Find the ?

What are the similarities between whole numbers and decimal numbers? ▷

\ Want to know /

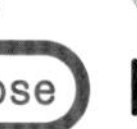

Purpose Is there any relationship between the structure of whole numbers and decimal numbers?

Decimal Numbers and Whole Numbers

Let's explore the structure and the size of numbers.

1 Let's explore about the numbers 1435 and 1.435.

① Let's write each number in the following table.

	Thousands	Hundreds	Tens	Ones	$\frac{1}{10}$	$\frac{1}{100}$	$\frac{1}{1000}$	
Elevation of Bakuchidake								m
Width of Shinkansen rail								m

Yu: I see that the 1 in 1435 represents 1000 and the 1 in 1.435 represents 1.

Sara: What is the value of each digit according to the place value?

② Let's fill in each ☐ with a number.

1435 is a number which gathers 1 set of ☐, 4 sets of ☐, 3 sets of ☐, and 5 sets of ☐.

1.435 is a number which gathers 1 set of ☐, 4 sets of ☐, 3 sets of ☐, and 5 sets of ☐.

Way to see and think
Think about how many sets you have for each place value.

③ Let's represent each number using math equations.

$1435 = 1000 + 400 + 30 + 5$

$= 1000 \times \square + 100 \times \square + 10 \times \square + 1 \times \square$

$1.435 = 1 + 0.4 + 0.03 + 0.005$

$= 1 \times \square + 0.1 \times \square + 0.01 \times \square + 0.001 \times \square$

1 Let's explore the structure of numbers. Let's compare and discuss the similarities between whole numbers and decimal numbers.

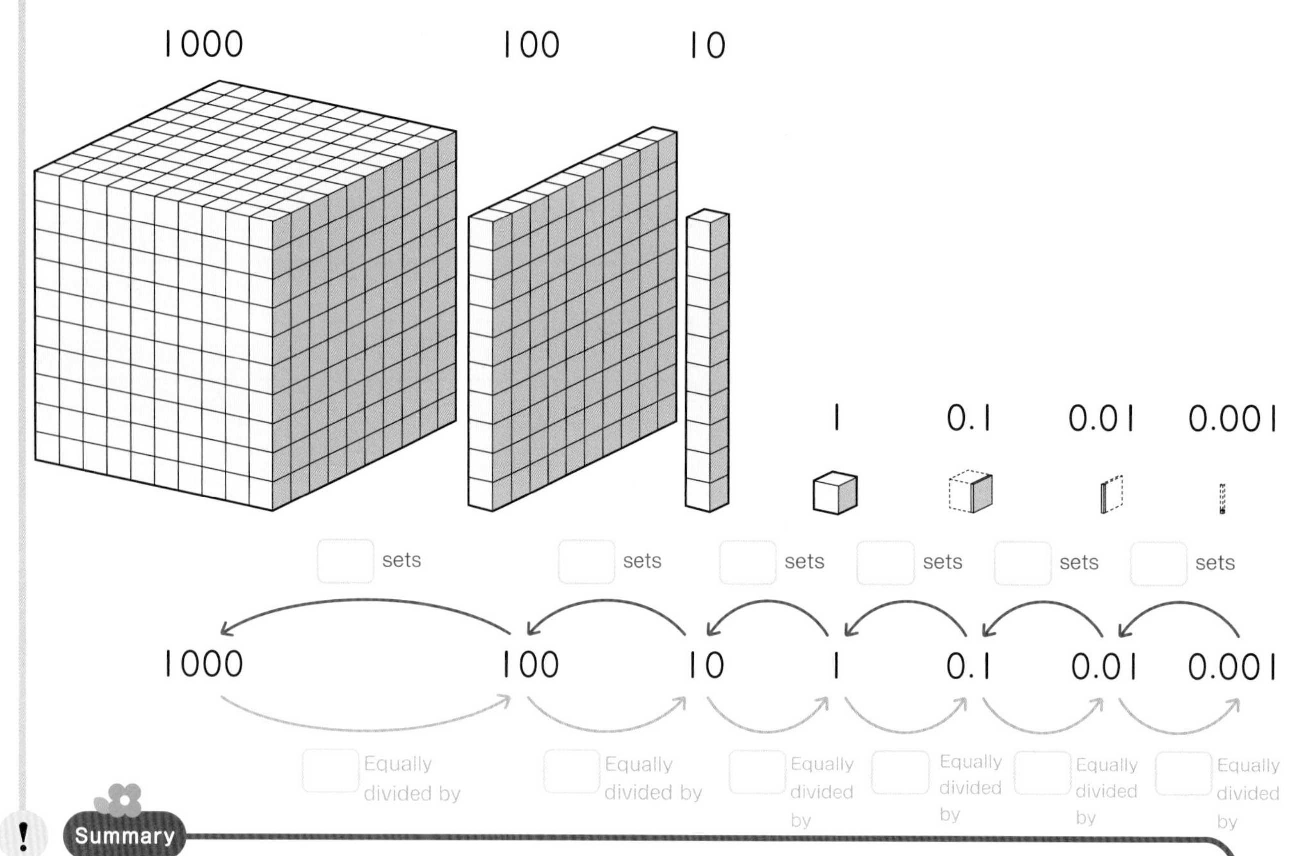

Summary

For both whole numbers and decimal numbers, if 10 sets are grouped then the number is moved to the next higher place value, and if the number is equally divided by 10 (equivalent to $\frac{1}{10}$) then the number is moved to the next lower place value.
Following this, any whole number or decimal number can be represented by using the ten numerals 0, 1, 2, ..., 9 and the decimal point.

2 Let's create the following numbers by using the ten numerals from 0 to 9 (only use once), and the decimal point.

① the smallest number

② the number that is less than 1 but closest to 1

? Since 1000 is 1000 times 1, can we say that 1435 is 1000 times 1.435?

2 Let's investigate about 10 times, 100 times, and 1000 times of 1.435.

With whole numbers, multiplying by 10 gives the number zeros to the right.

If we consider the case of decimal numbers in the same way...

Want to explore

Purpose What number will it be when you calculate 10 times, 100 times, and 1000 times of a number?

① Let's find 10 times, 100 times, and 1000 times of 1.435, and write it down in the following table.

	Thousands	Hundreds	Tens	Ones	$\frac{1}{10}$	$\frac{1}{100}$	$\frac{1}{1000}$
				1.	4	3	5
10 times of 1.435 →							
100 times of 1.435 →							
1000 times of 1.435 →							

(10 times, 10 times, 10 times; 100 times; 1000 times)

② How do the decimal places change when multiplied by 10, 100, or 1,000?

③ Let's fill in the ☐ below with the decimal point when you calculate 10 times, 100 times, and 1000 times of a number.

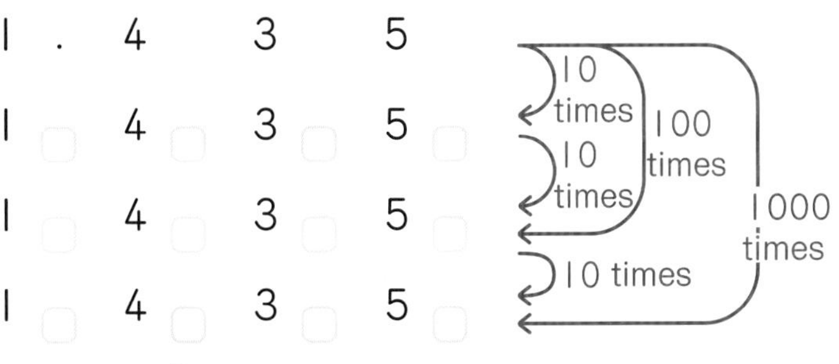

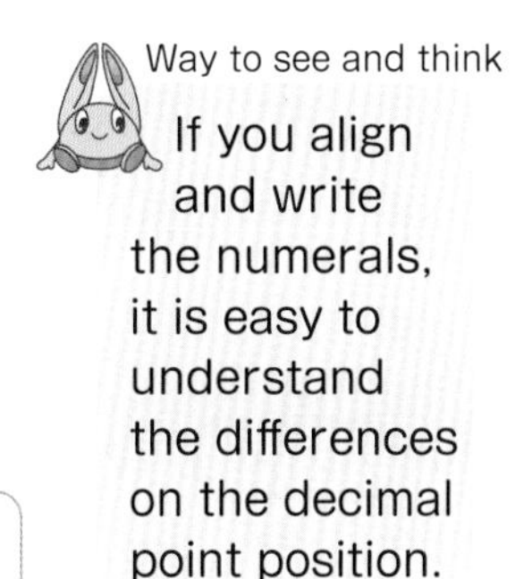

Way to see and think
If you align and write the numerals, it is easy to understand the differences on the decimal point position.

When you multiply a decimal by 10, 100, or 1000, the position of the decimal point changes.

Summary

When you identify 10 times, 100 times, 1000 times,... of a number, the number's decimal point moves respectively 1 place, 2 places, 3 paces, ..., to the right.

4 Let's try to represent with math equations 10 times, 100 times, and 1000 times of 1.435

ⓐ $1.435 \times 10 =$ ☐

ⓑ $1.435 \times 100 =$ ☐

ⓒ $1.435 \times 1000 =$ ☐

1 Let's answer the following questions.

① Let's write 10 times, 100 times, and 1000 times of 23.47.

② How many times of 8.72 do 87.2, 872, and 8720 represent, respectively?

? Is there a rule that the number can be written as a set of $\frac{1}{10}$?

3 Let's explore about $\frac{1}{10}$ and $\frac{1}{100}$ of 125.

If you have $\frac{1}{10}$ of a number and there is a zero on the right-hand side, the zero is eliminated.

If the rightmost digit is not zero, then....

\ Want to explore /

? **Purpose** What number will it be when you calculate $\frac{1}{10}$ and $\frac{1}{100}$ of a number?

① Let's find $\frac{1}{10}$ and $\frac{1}{100}$ of 125, and write it down in the following table.

	Hundreds	Tens	Ones	$\frac{1}{10}$	$\frac{1}{100}$
	1	2	5		
$\frac{1}{10}$ of 125 →					
$\frac{1}{100}$ of 125 →					

As for the number with size $\frac{1}{10}$ of 125, since:
$100 \div 10 = 10$,
$20 \div 10 = 2$,
$5 \div 10 = 0.5$ and
$10 + 2 + 0.5 = 12.5$,
then the answer is 12.5.

② If you calculate $\frac{1}{10}$ and $\frac{1}{100}$ of a number, how does the place value of the number change?

③ Let's fill in the ☐ below with the decimal point when you calculate $\frac{1}{10}$ and $\frac{1}{100}$ of 125.

1 2 5

1 ☐ 2 ☐ 5 ☐

1 ☐ 2 ☐ 5 ☐

Way to see and think

If you align and write the numerals, it is easy to understand the differences on the decimal point position.

Summary

When you identify $\frac{1}{10}$, $\frac{1}{100}$, ..., of a number, the number's decimal point moves respectively 1 place, 2 places, ..., to the left.

④ Let's try to represent with math equations $\frac{1}{10}$ and $\frac{1}{100}$ of a number.

ⓐ $125 \div 10 =$ ☐

ⓑ $125 \div 100 =$ ☐

⑤ What number is $\frac{1}{1000}$ of 125?

1 Let's answer the following questions.

① Let's write $\frac{1}{10}$ and $\frac{1}{100}$ of 30.84.

② What fraction of 63.2 do 6.32 and 0.632 represent respectively?

CAN What can you do?

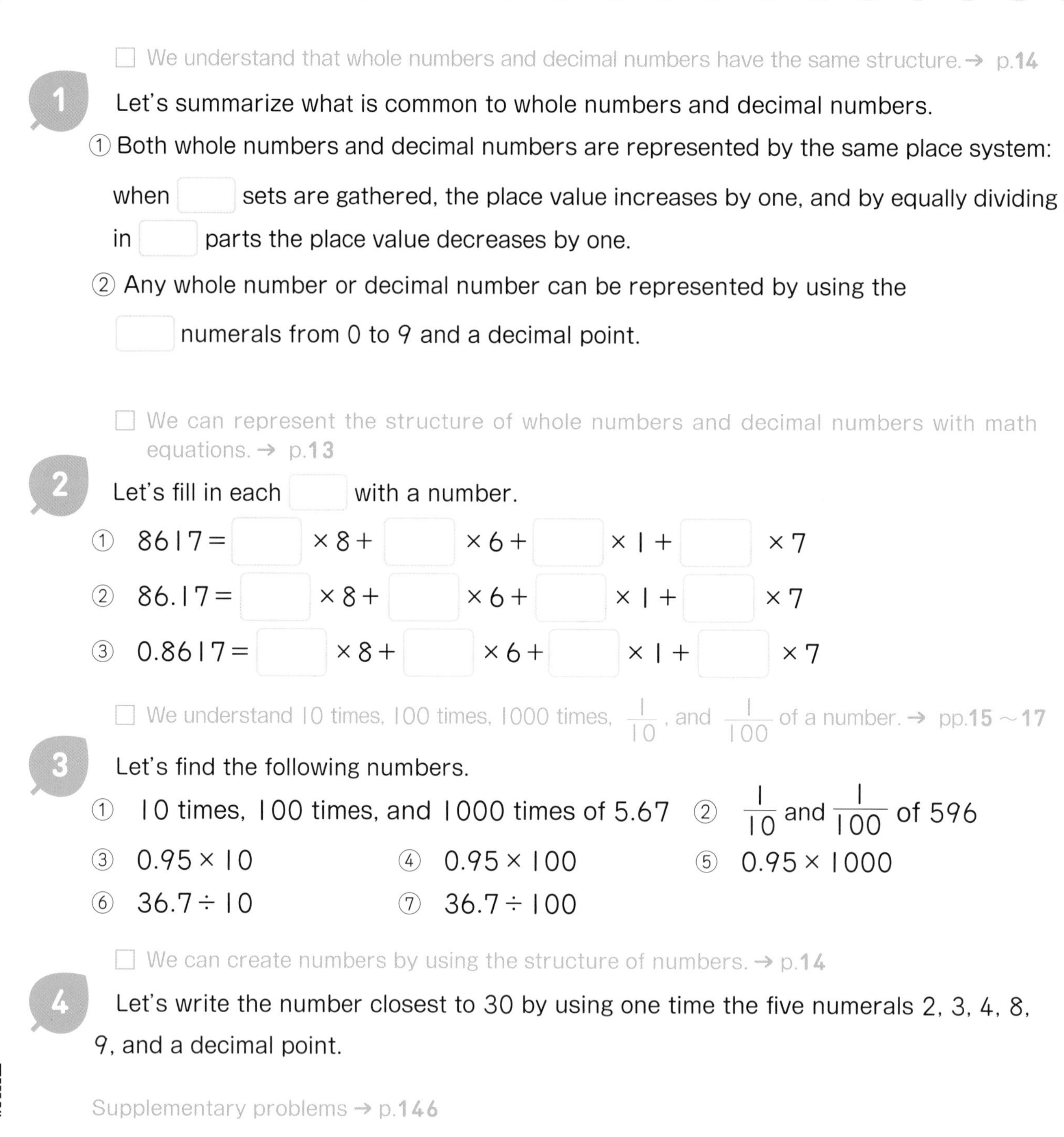

☐ We understand that whole numbers and decimal numbers have the same structure. → p.14

1 Let's summarize what is common to whole numbers and decimal numbers.

① Both whole numbers and decimal numbers are represented by the same place system: when ☐ sets are gathered, the place value increases by one, and by equally dividing in ☐ parts the place value decreases by one.

② Any whole number or decimal number can be represented by using the ☐ numerals from 0 to 9 and a decimal point.

☐ We can represent the structure of whole numbers and decimal numbers with math equations. → p.13

2 Let's fill in each ☐ with a number.

① $8617 = \square \times 8 + \square \times 6 + \square \times 1 + \square \times 7$

② $86.17 = \square \times 8 + \square \times 6 + \square \times 1 + \square \times 7$

③ $0.8617 = \square \times 8 + \square \times 6 + \square \times 1 + \square \times 7$

☐ We understand 10 times, 100 times, 1000 times, $\frac{1}{10}$, and $\frac{1}{100}$ of a number. → pp.15 ~ 17

3 Let's find the following numbers.

① 10 times, 100 times, and 1000 times of 5.67　② $\frac{1}{10}$ and $\frac{1}{100}$ of 596

③ 0.95×10　④ 0.95×100　⑤ 0.95×1000

⑥ $36.7 \div 10$　⑦ $36.7 \div 100$

☐ We can create numbers by using the structure of numbers. → p.14

4 Let's write the number closest to 30 by using one time the five numerals 2, 3, 4, 8, 9, and a decimal point.

Supplementary problems → p.146

Which "Way to See and Think Monsters" did you find in "1 Decimal Numbers and Whole Numbers"?

I found "Unit" when I represented various numbers.

Haruto

Also, when decimal numbers are multiplied by 10 or $\frac{1}{10}$ ...

Akari

With the Way to See and Think Monsters...

Let's Reflect!

Let's reflect on which monster you used while learning "1 Decimal Numbers and Whole Numbers."

Unit

Considering each place as one unit, we found out that we can represent any whole number or decimal number can be represented by using the numerals from 0 to 9 and a decimal point.

① What are the structures of decimal numbers and whole numbers?

$$38.05 = 10 \times \square + 1 \times \square + 0.1 \times \square + 0.01 \times \square$$

Akari

38.05 is the number that gathers 3 sets of 10, 8 sets of 1, no set of 0.1, and 5 sets of 0.01.

Haruto

Both decimal and whole numbers can be viewed as how many numbers there are in each place.

Rule

There is a rule that a number multiplied by 10 is a number with the decimal point of the original number shifted one place to the right, and that a number multiplied by $\frac{1}{10}$ is a number with the decimal point of the original number shifted one place to the left.

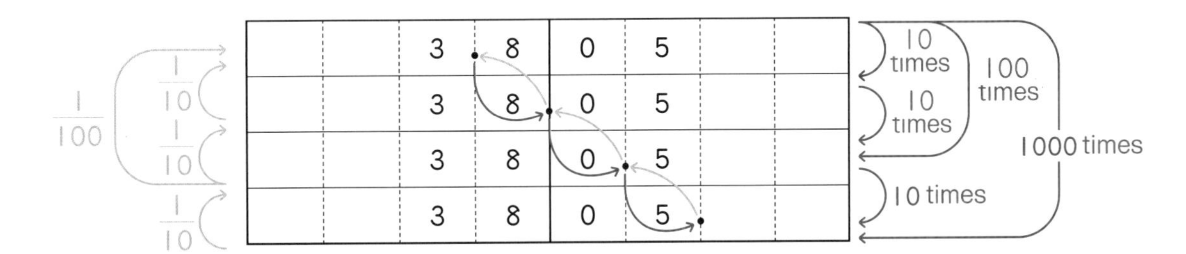

Solve the ?

We found out that decimal numbers and whole numbers are formed by the same structure.

Sara

Want to Connect

Can we do various calculations using the number structures we have learned so far?

Yu

Find the ?

Which one is the same?

Sara is doing a tangram puzzle.
1
Look, look, there are more puzzles with these figures.
2
Ah!
3
Ⓐ
Ⓑ
Ⓒ
Ⓓ
Ⓔ
Ⓕ
Ⓖ
Ⓗ
Ⓘ
Ⓙ
Our figures were mixed.
4
Which are mine?
It would be easy if I could find the shapes that overlap exactly...
5

\ Want to know /

Purpose How can we find the shapes that overlap exactly?

2 Congruent Figures

Let's explore the properties of figures with the same shape and size and how to draw them.

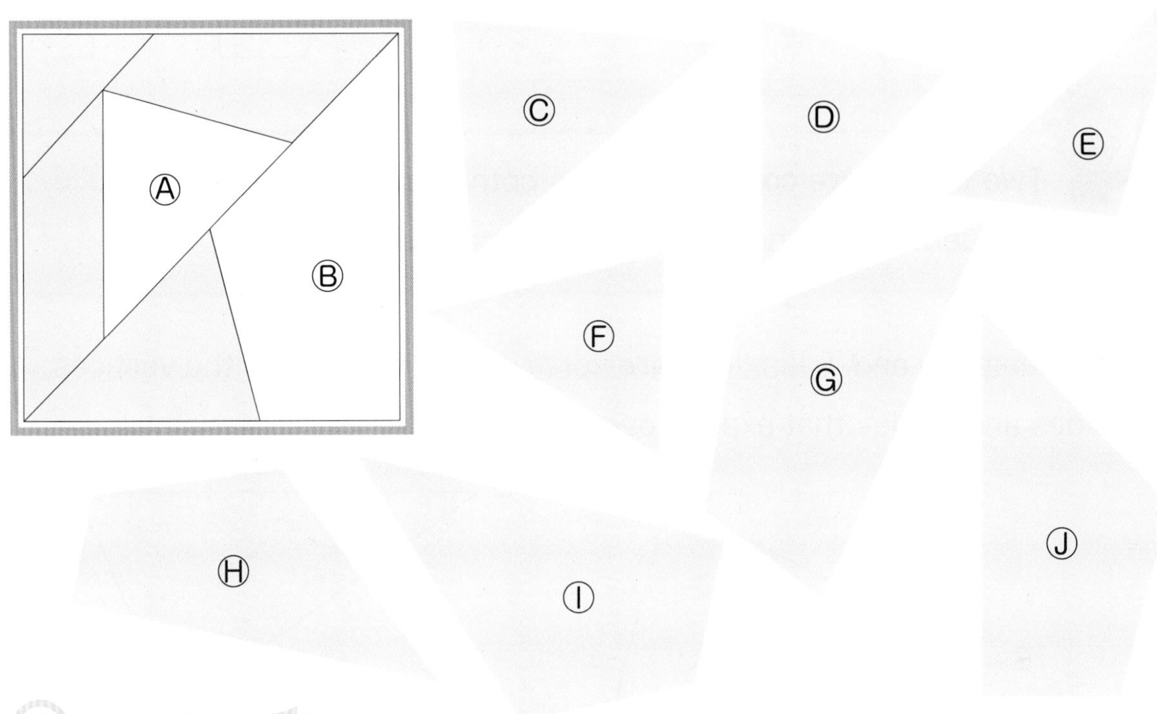

1 Congruent Figures

Exactly overlapping figures →

1 Let's find the figures that exactly overlap Figure Ⓐ and Figure Ⓑ, respectively.

By only looking at it, you can't immediately know which one.

Which part should I look at to find that the size and shape of two figures are the same?

① Let's find a figure from Ⓒ~Ⓕ that exactly overlaps with triangle Ⓐ.

② Let's find a figure from Ⓖ~Ⓙ that exactly overlaps with quadrilateral Ⓑ.

Let's cut out the figures on page 173 and try to explore.

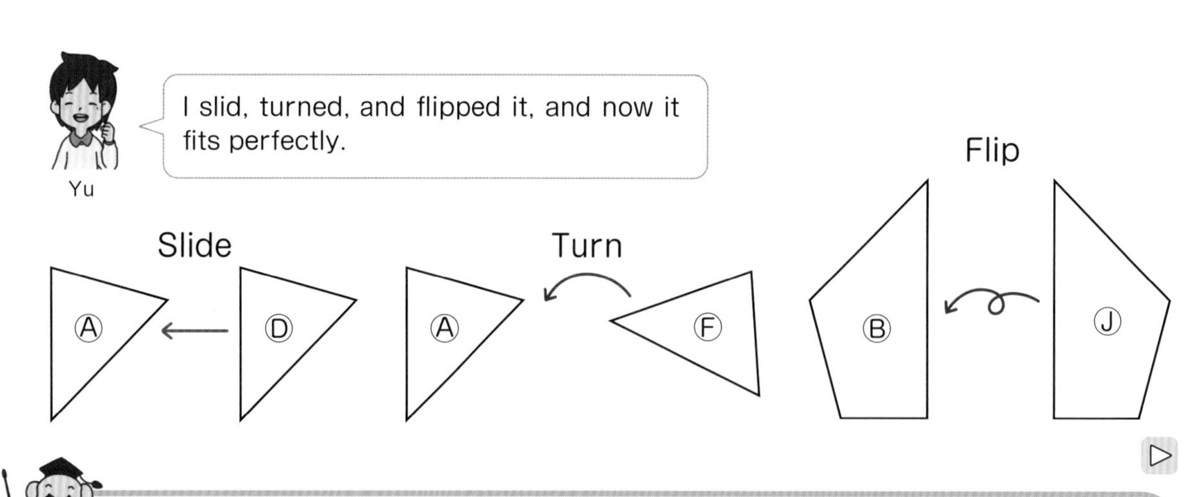

I slid, turned, and flipped it, and now it fits perfectly.
Yu
Slide
Turn
Flip
Ⓐ
Ⓓ
Ⓐ
Ⓕ
Ⓑ
Ⓙ

Two figures are **congruent** when both exactly overlap.
Congruent figures have the same shape and size.

1 Triangle Ⓚ and Triangle Ⓛ are congruent. Let's find all the vertices, sides and angles that exactly overlap.

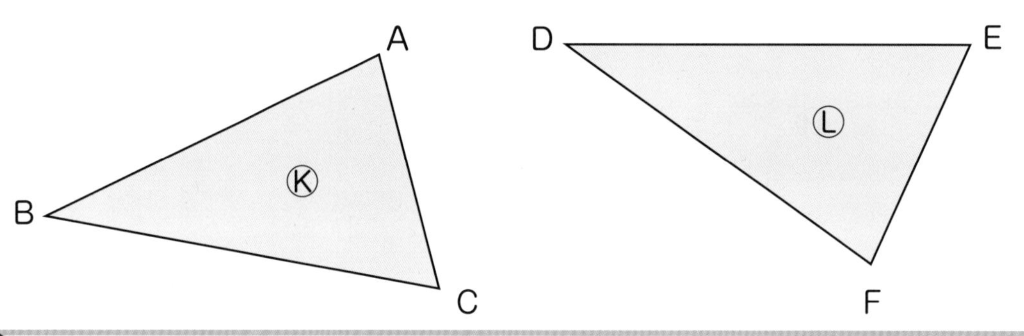

In congruent figures, overlapping vertices, overlapping sides, and overlapping angles are called **corresponding vertices, corresponding sides,** and **corresponding angles** respectively.

Summary

By sliding, turning, and flipping, you can find shapes that overlap exactly.

? In congruent figures, what are the properties of the corresponding side lengths and the corresponding angle sizes?

2

The quadrilaterals Ⓜ and Ⓝ, shown on the right, are congruent. Let's compare the length of the corresponding sides and the size of the corresponding angles. What can you see?

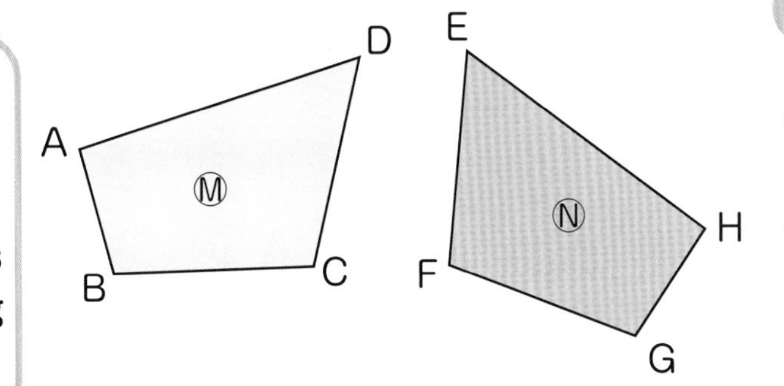

Want to explore

? **Purpose** What are the properties of congruent figures?

1. Compare the lengths of corresponding sides
2. Compare the size of corresponding angles.

1 Let's compare the length of the corresponding sides and the size of the corresponding angles of the triangles Ⓚ and Ⓛ in **1** of the previous page.

! **Summary**

In congruent figures, the corresponding sides have equal length. Also, the corresponding angles have equal size.

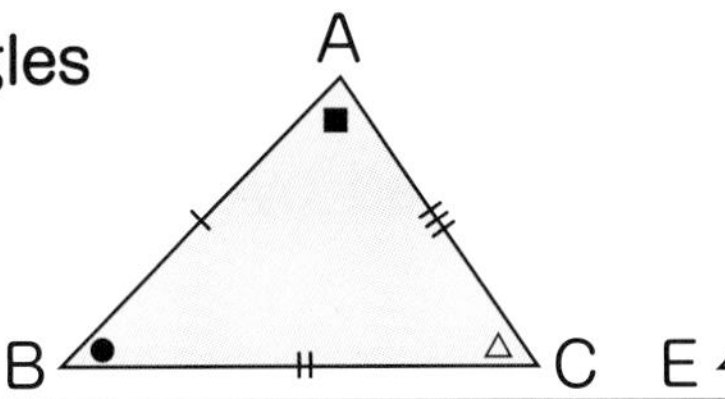

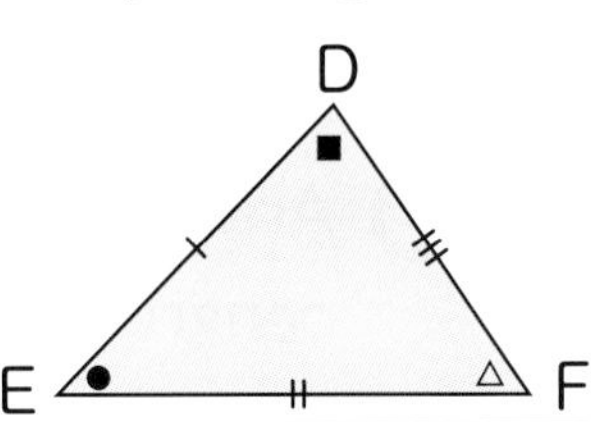

2 The triangles Ⓐ and Ⓑ, shown on the right, are congruent. Let's explore about corresponding sides and angles.

① Which side is corresponding to side DF?
What is the length in cm of the side DF?

② Which angle is corresponding to angle E?
What is the size in degrees of angle E?

③ Let's find the length of the other sides and the size of the other angles in triangle Ⓑ.

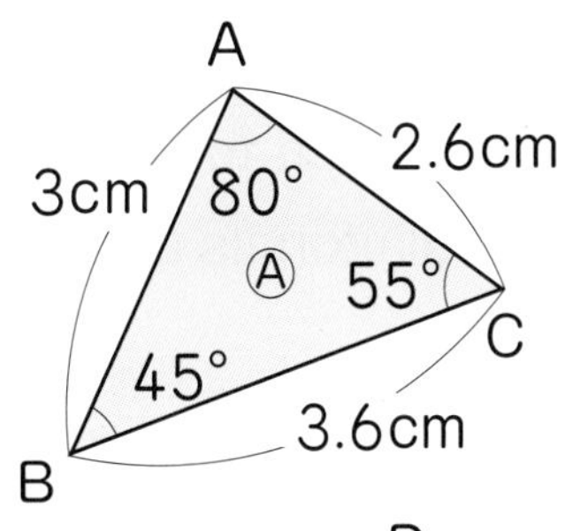

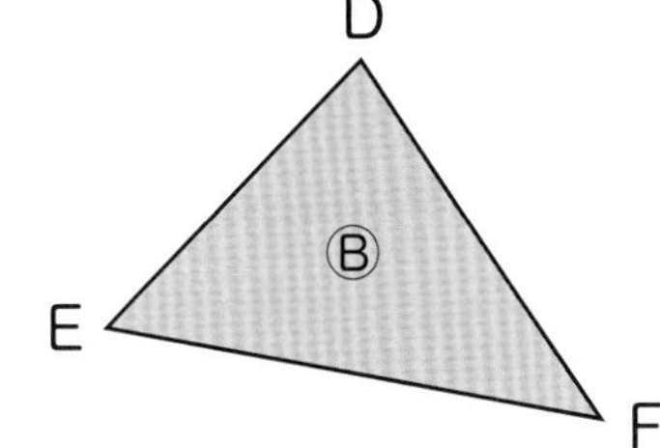

3 Are quadrilaterals Ⓒ and Ⓓ congruent? If so, why?

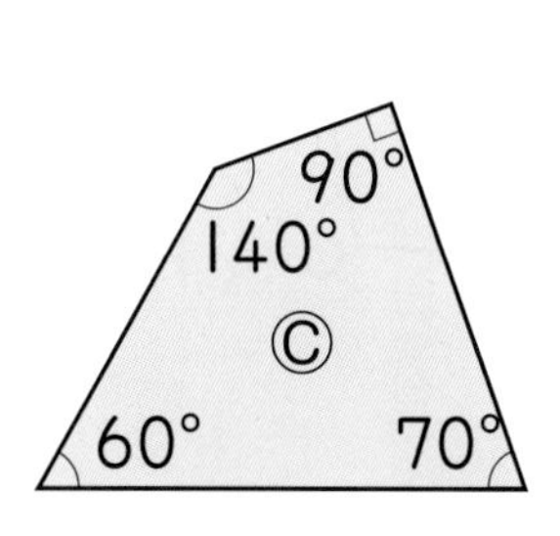

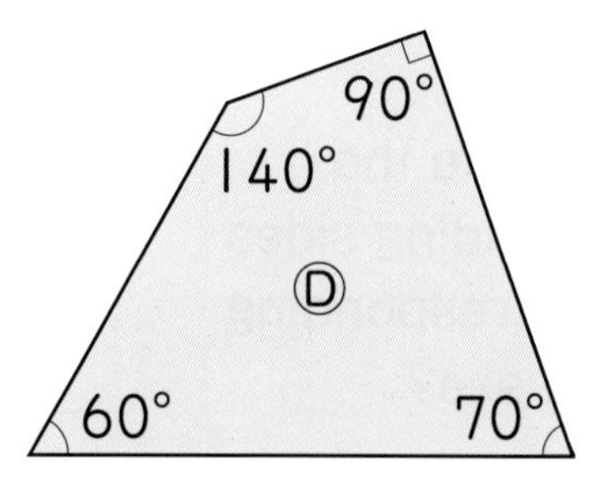

Way to see and think

Which properties of congruent figures can you identify?

? If you divide a quadrilateral by a diagonal line, you get two triangles, but are they congruent?

Shapes divided by diagonals →

3

Consider the triangles formed by drawing a diagonal line through the following quadrilaterals.

ⓐ Trapezoid ⓑ Parallelogram ⓒ Square ⓓ Rectangle ⓔ Rhombus

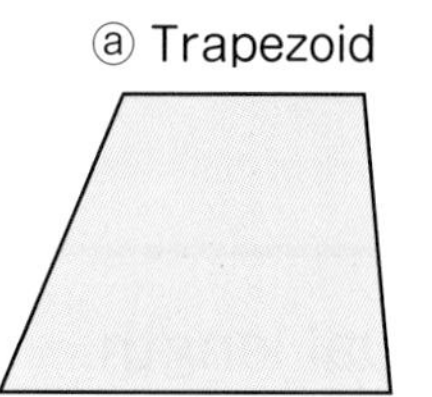

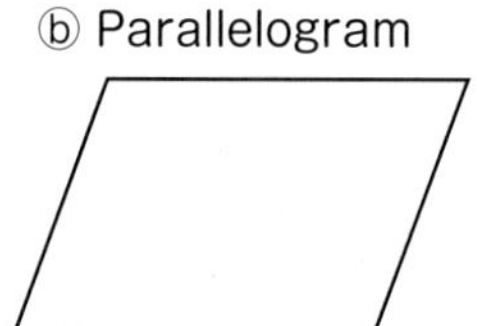

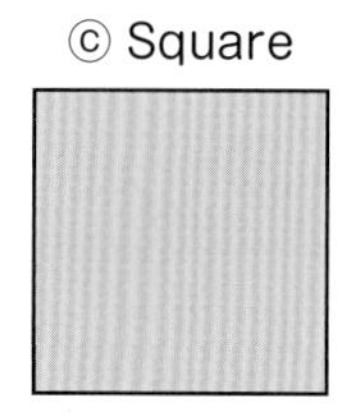

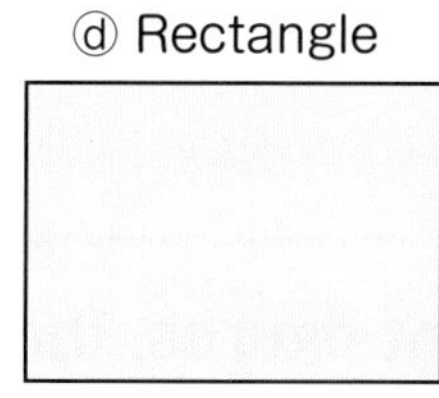

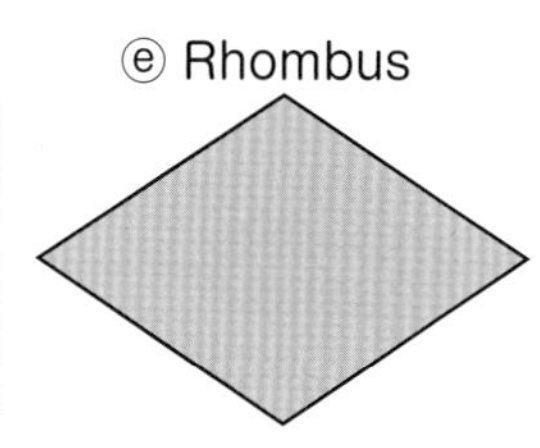

\ Want to know /

? **Purpose** **Are triangles divided by a diagonal line of a quadrilateral always congruent?**

① Which two triangles made by drawing one diagonal are congruent?

② Which four triangles made by drawing two diagonals are congruent?

Is there a quadrilateral where all four triangles are congruent?

! **Summary**

Some quadrilaterals can be viewed as a combination of congruent triangles.

? Can we use what we have found so far to create congruent figures?

2 How to draw congruent figures?

1 Let's think about how to draw a triangle that is congruent to the triangle shown on the right.

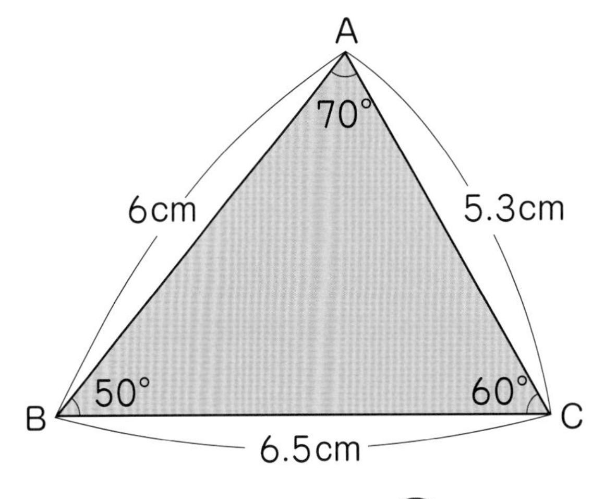

All you have to know is the position of vertices A, B, and C.

What lengths and angles could we use?

Want to know

Purpose How can we draw a congruent triangle?

① Let's think about how to draw by using a compass and protractor.

I drew a straight line with the same length as side BC.

After, all you need to know is the position of vertex A.

② Let's discuss how to determine the position of vertex A.

Sara's drawing method

Measure the length of the 3 sides, and then draw them.

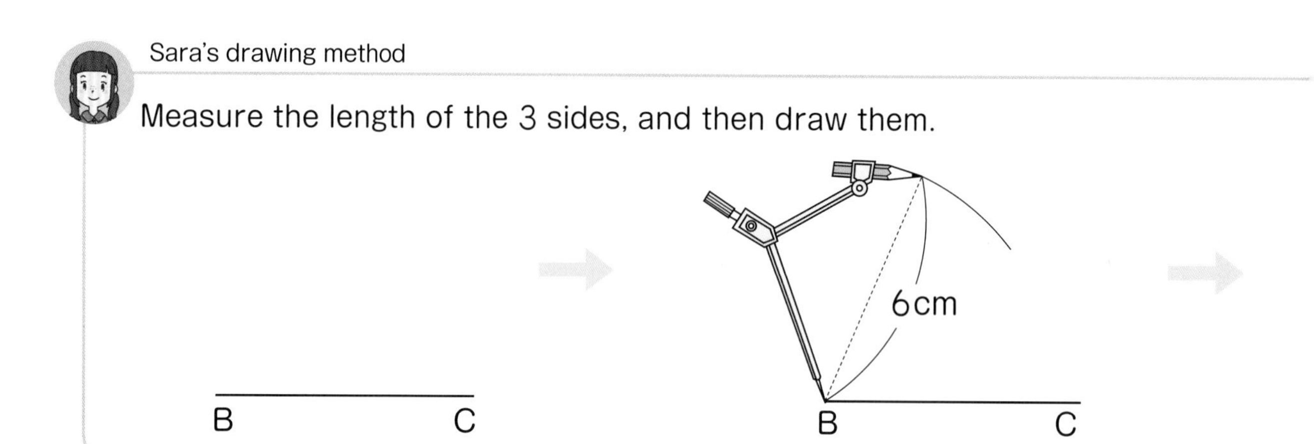

Akari's drawing method

Measure the length of 2 sides and the angle in between, and then draw them.

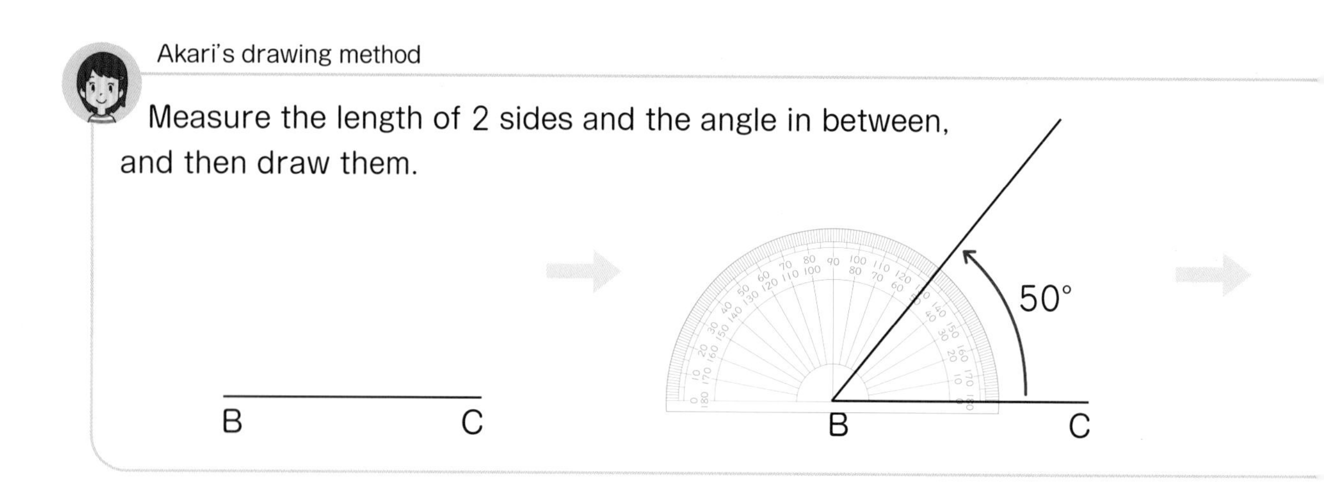

Yu's drawing method

Measure the length of 1 side and the angles formed by that side with the other two sides, and then draw them.

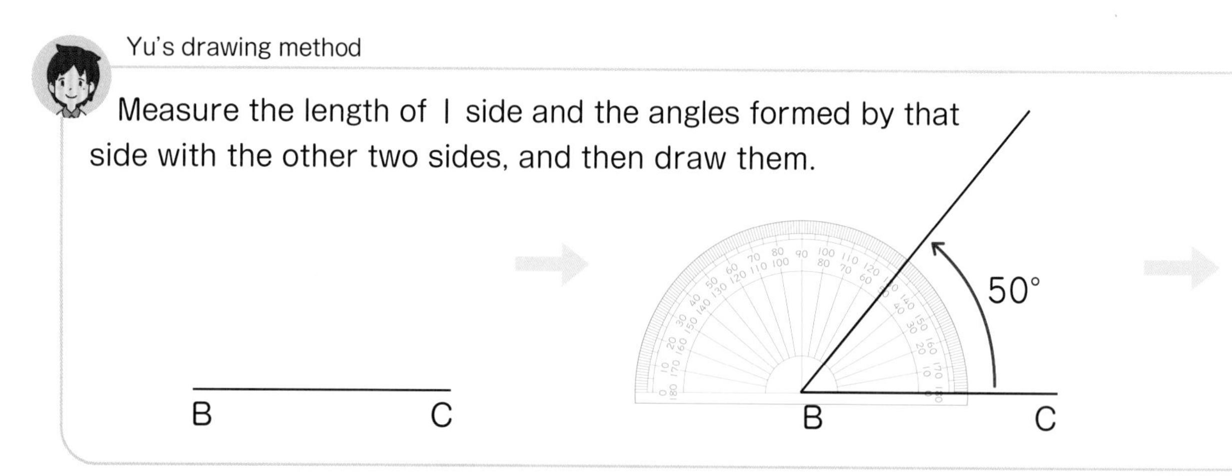

③ Let's explain the process followed by the 3 children on each drawing method.

④ Let's confirm that the triangle ABC drawn is congruent to the triangle from 1 in the previous page.

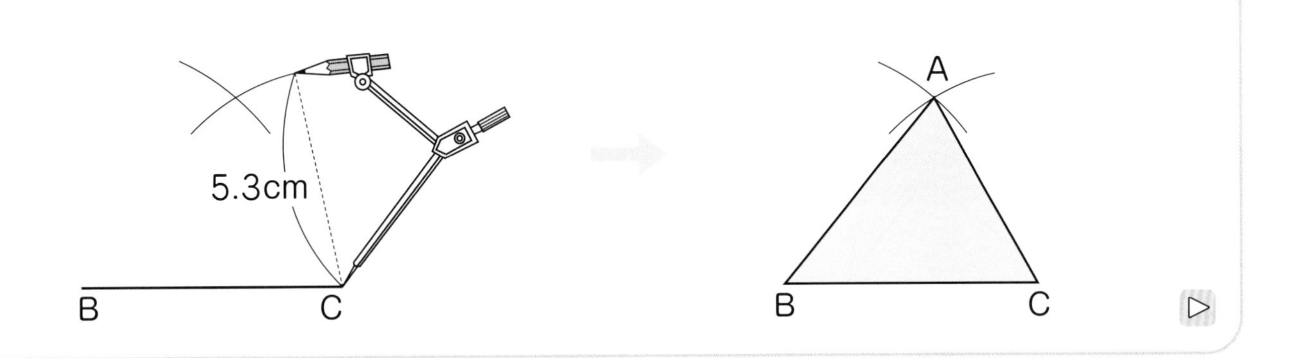

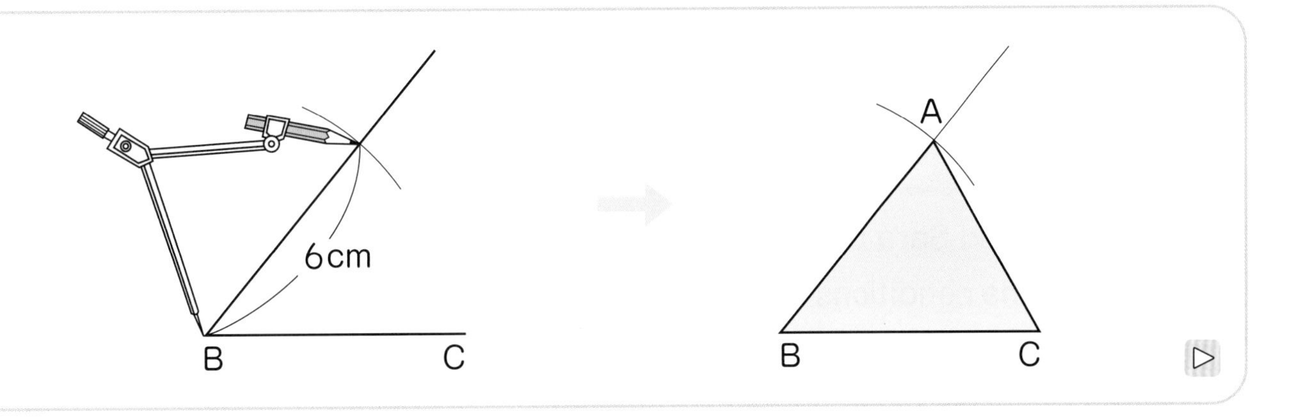

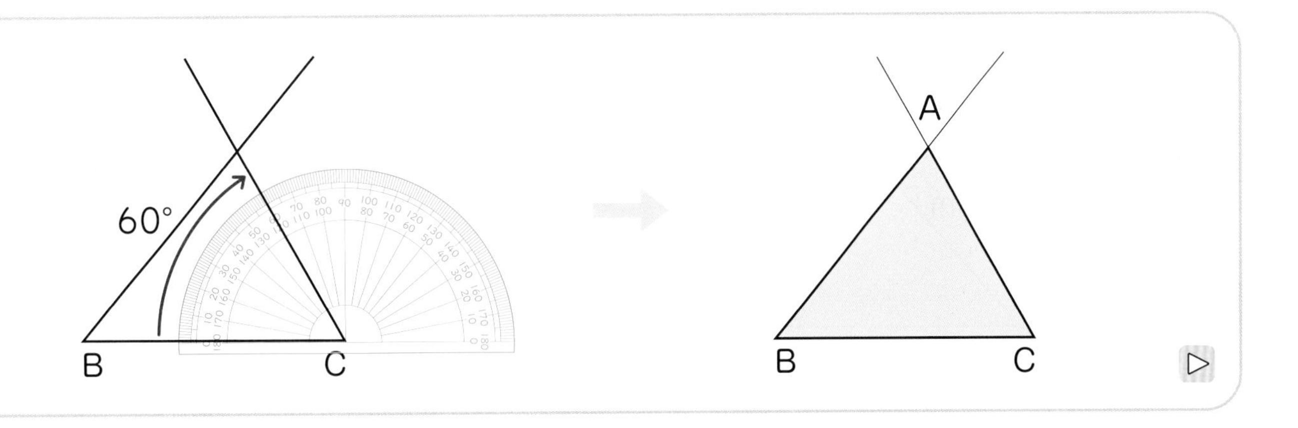

Do not erase the lines you use in the drawing process.

! **Summary**

Among the length of the 3 sides and the size of the 3 angles, if one of the following Ⓐ, Ⓑ or Ⓒ is known, then a congruent triangle can be drawn.

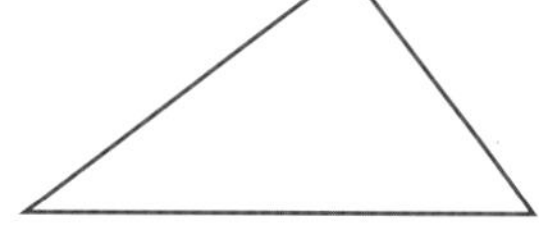

Ⓐ Length of the 3 sides.

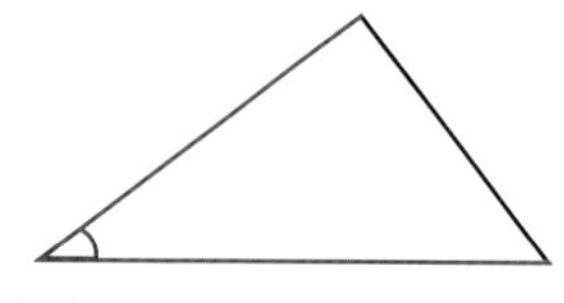

Ⓑ Length of 2 sides and the angle in between.

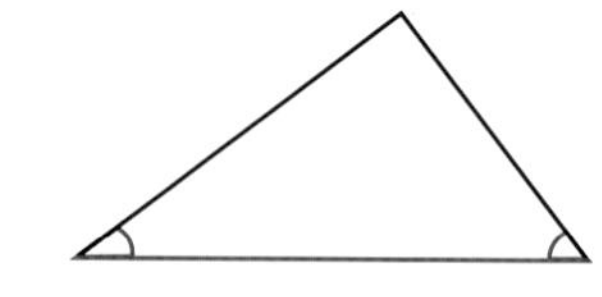

Ⓒ Length of one side and the angles formed by that side with the other two sides.

? Can we draw a congruent figure if we know at least three of the side lengths and angle sizes?

Reasons for not having congruent triangles →

2

Haruto and Sara drew triangle ABC under the conditions shown on the right. Let's discuss the reasons why the drawings from the 2 children are different.

Length of side AB is 4 cm.
Length of side BC is 4.4 cm.
Size of angle C is 60°.

Haruto's triangle

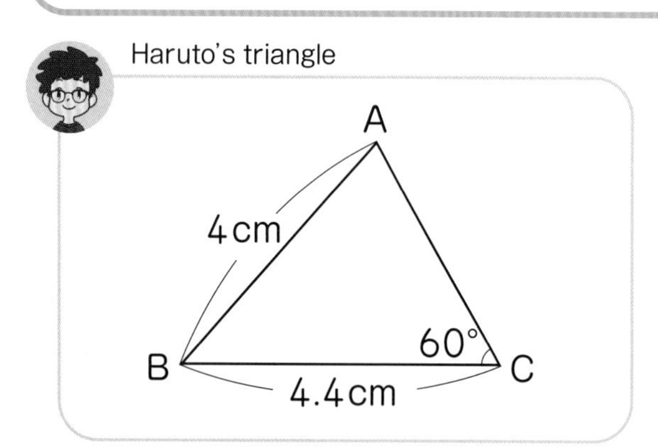

Sara's triangle

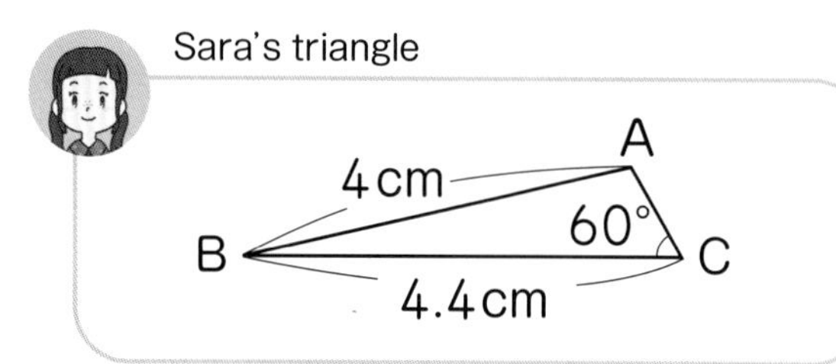

\ Want to think /

Purpose Let's try to find the reasons why this happens.

Sara

Using the diagram, we can think of it as...

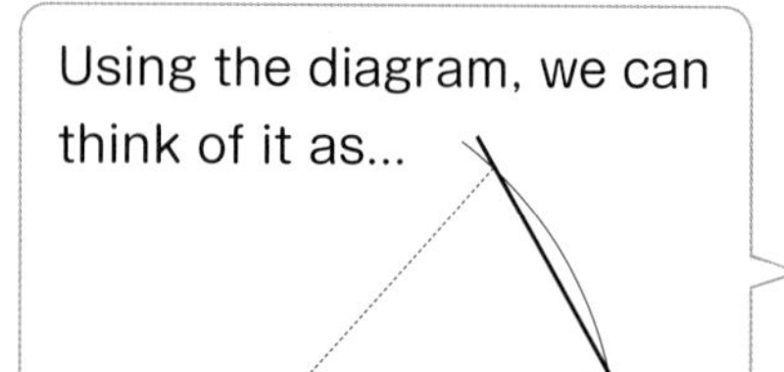

Akari

I wonder if you can draw a congruent triangle if you only know the size of the 3 angles.

Haruto

1 Draw a congruent triangle to the triangle shown on the right, by measuring the required length of the sides or size of the angles. Let's explain how you drew it.

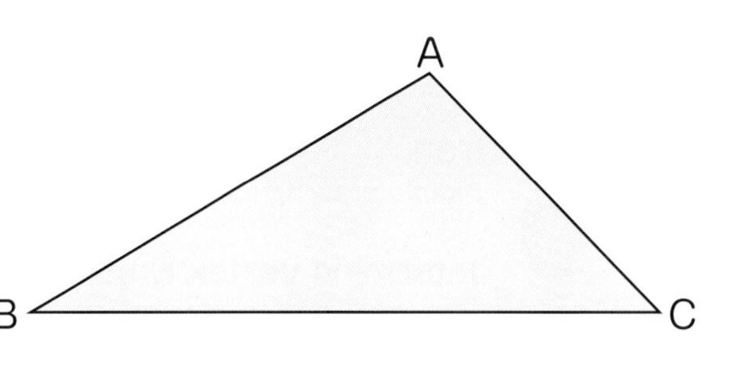

? Can we also draw congruent quadrilaterals?

3 **Let's think about how to draw a quadrilateral that is congruent to the following quadrilateral.**

Haruto: Same as with the triangle, can I draw it if I know the length of the 4 sides?

Sara: I wonder if I can use the drawing methods for congruent triangles?

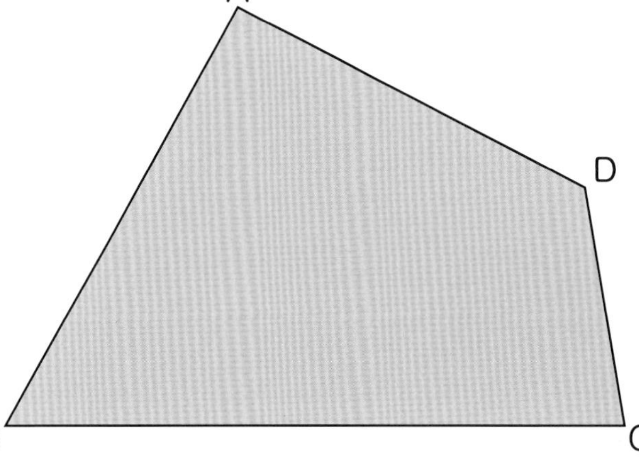

\ Want to think /

? **Purpose** **How can we draw a congruent quadrilateral?**

① Can you draw a congruent quadrilateral only using the length of the 4 sides?

② Let's measure the length of the sides and the size of the angles from the figures above, and draw congruent quadrilaterals in your notebook.

③ Let's explain the drawing methods of the following children.

Way to see and think

You are thinking based on the drawing methods for congruent triangles.

Akari's drawing method

Determine vertex D using the idea from triangles. Measure the length of side AD and side CD.

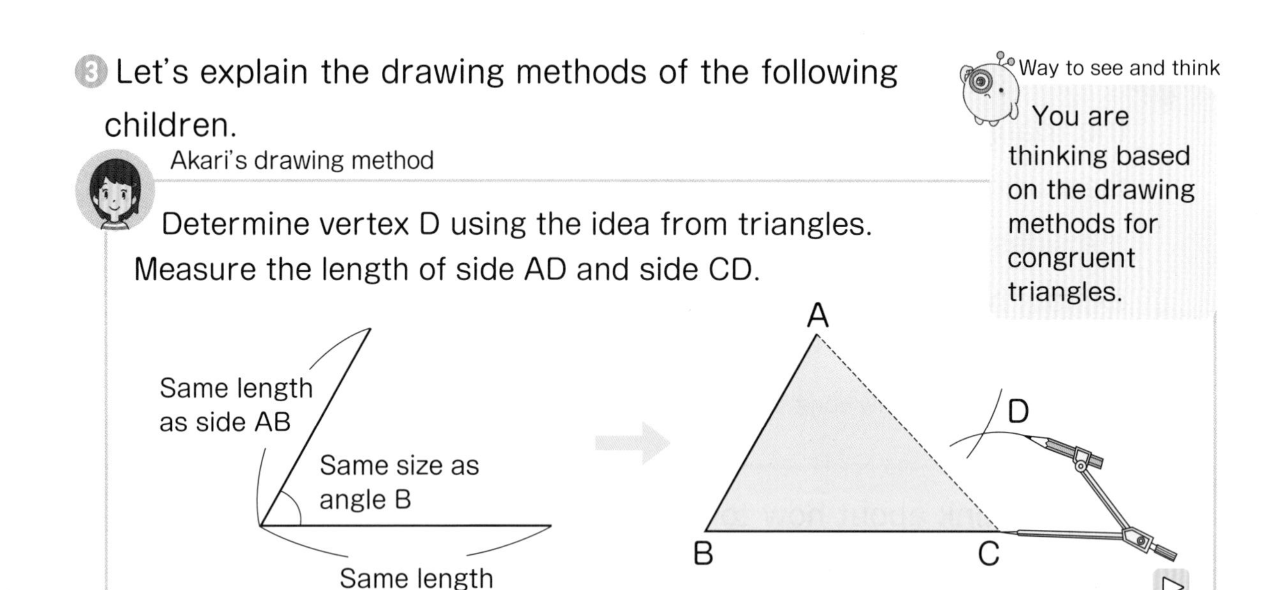

Yu's drawing method

Determine vertex D using the idea from triangles. Measure the angles formed by diagonal AC and the other two sides.

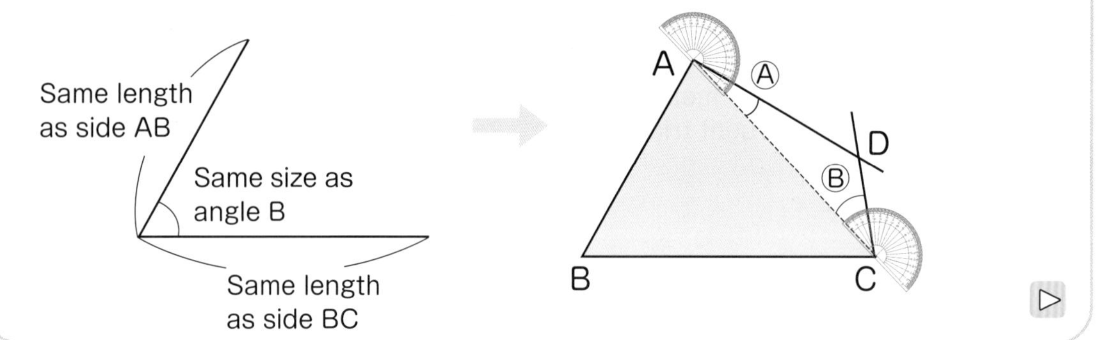

Summary — Way to see and think

Congruent quadrilaterals can be drawn by using the drawing methods from congruent triangles if the quadrilateral is divided into two triangles by a diagonal.

1 Let's draw a quadrilateral that is congruent to the quadrilateral shown on the right.

Which sides and angles should we use?

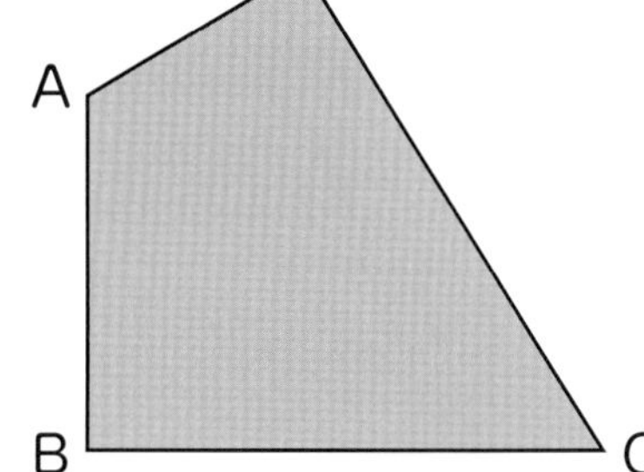

4 Let's tessellate congruent triangles like the diagram shown below. Also, let's discuss what you notice.

\ Want to think /

Purpose Let's focus on the corresponding angles of congruent triangles.

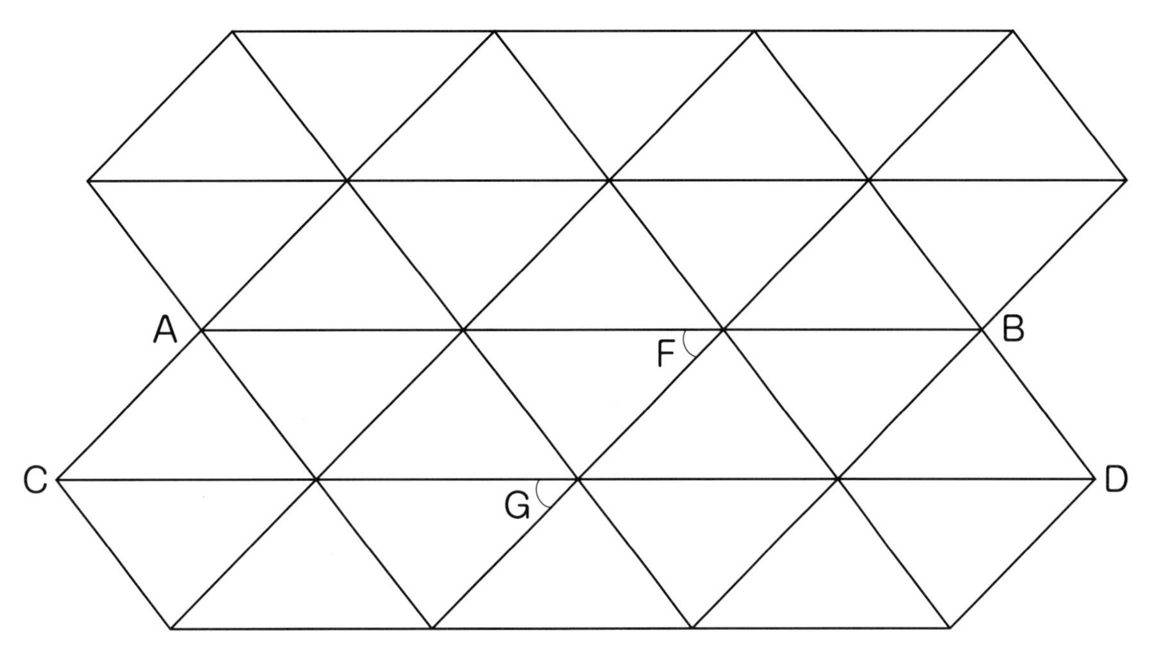

① Is the size of angle F and angle G equal?

② Verify that straight line AB and straight line CD are parallel, use triangle rulers.

③ Let's find parallelograms in the diagram above, and explain how each become a parallelogram.

④ Let's find trapezoids in the diagram above and explain how each become a trapezoid by using the words "parallel" and "straight line."

Way to see and think

Clarify the reason and explain in order.

What other shapes can you find?

CAN What can you do?

☐ We understand corresponding vertices, sides and angles of congruent figures. → p.23

1 The two quadrilaterals on the right are congruent. Let's say the length of all the corresponding sides and the size of all the corresponding angles.

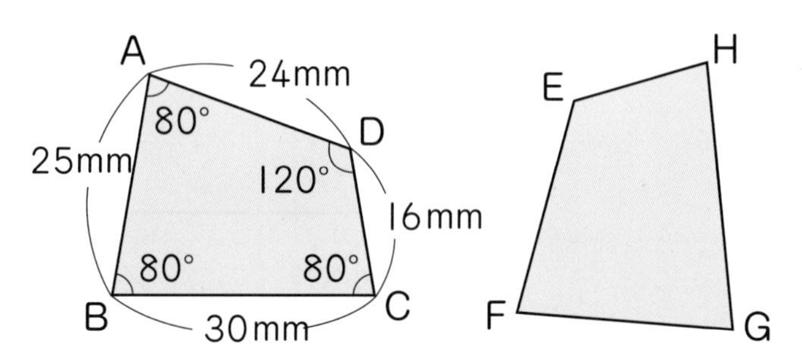

☐ We can draw congruent triangles. → pp.25 ~ 27

2 Let's draw a triangle that is congruent to the following triangles.

① A triangle with side lengths of 4 cm, 7 cm, and 8 cm.

② A triangle with two angles of size 45° and 60°, and the side in between with length of 6 cm.

③

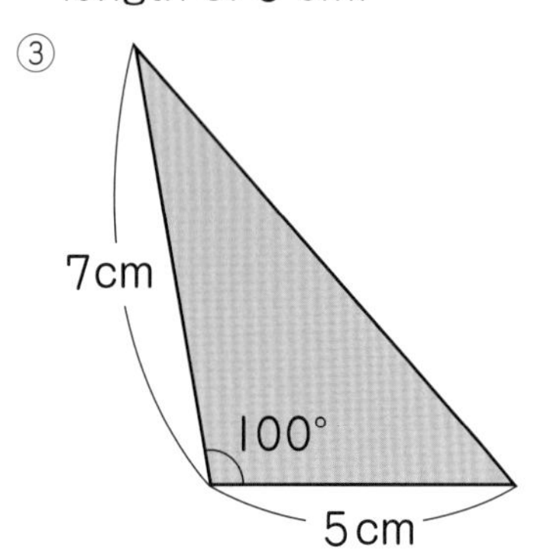

④

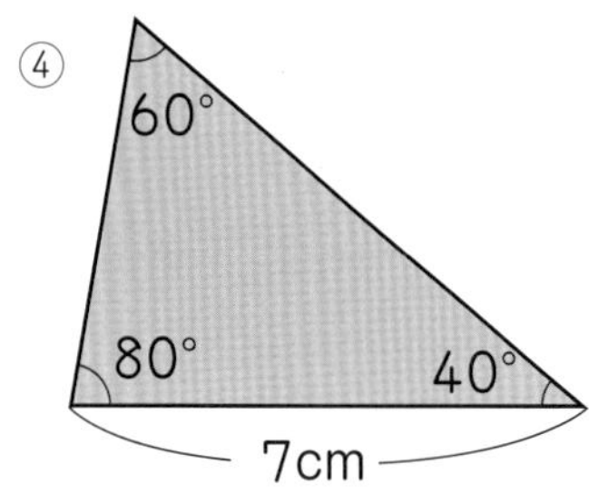

☐ We can draw congruent quadrilaterals. → p.30

3 Let's draw a rhombus that is congruent to the rhombus on the right.

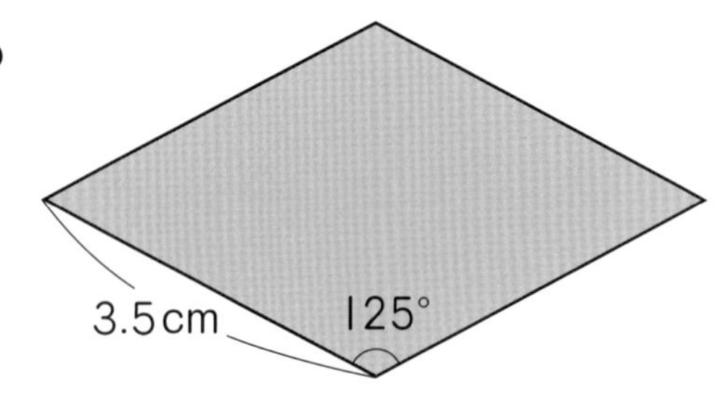

Supplementary problems → p.157

Which "Way to See and Think Monsters" did you find in "2 Congruent Figures"?

I found "Why" when explaining congruent figures.

Sara

I found other monsters, too!

Yu

Utilize Usefulness and Efficiency of Learning

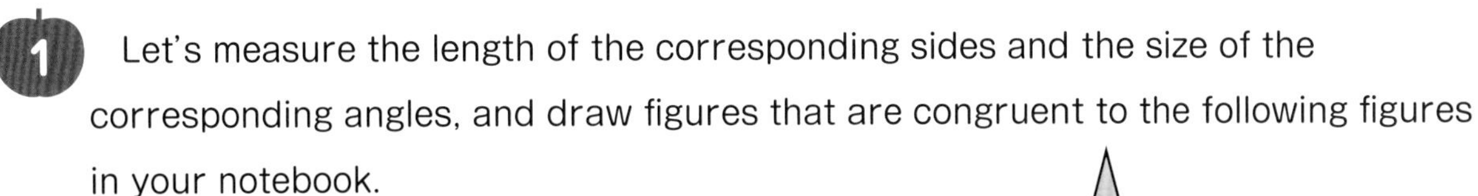

1 Let's measure the length of the corresponding sides and the size of the corresponding angles, and draw figures that are congruent to the following figures in your notebook.

① Equilateral Triangle

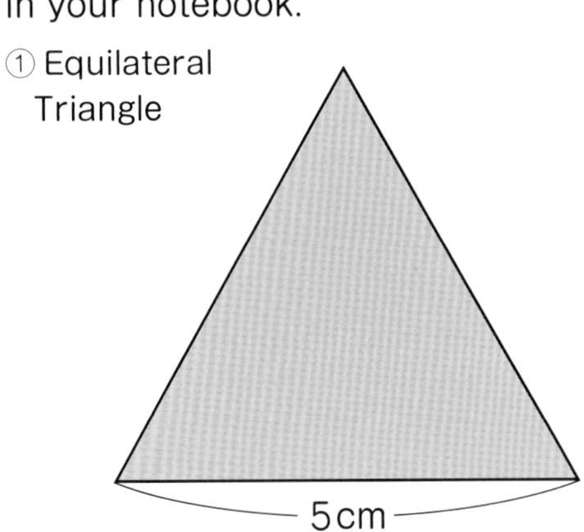

② Isosceles Triangle

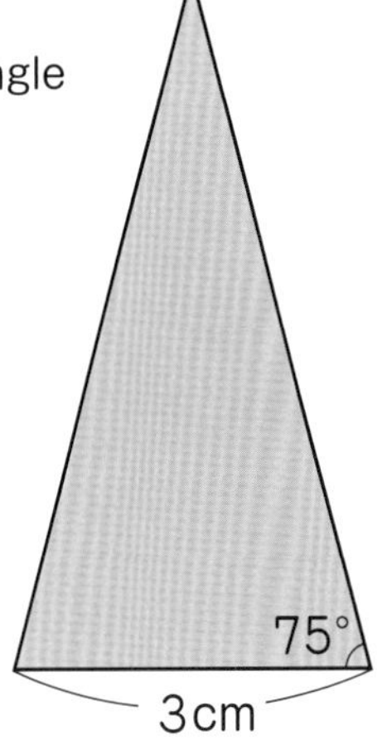

2 In the question **1** Akari thought as follows. Let's discuss why she thought so.

Akari's idea

For the equilateral triangle in ①, we can draw congruent triangles without measuring the lengths or the sizes of the angles, just with noticing that the length of one side is 5 cm is enough.

The properties of equilateral triangles are ...

I wonder if the isosceles triangle in ② will look the same way.

3 There are several triangular rulers like the one on the right. Solve the following situations.

① Let's make a rectangle or a parallelogram using 2 rulers.

② Use 4 rulers to make a rhombus.

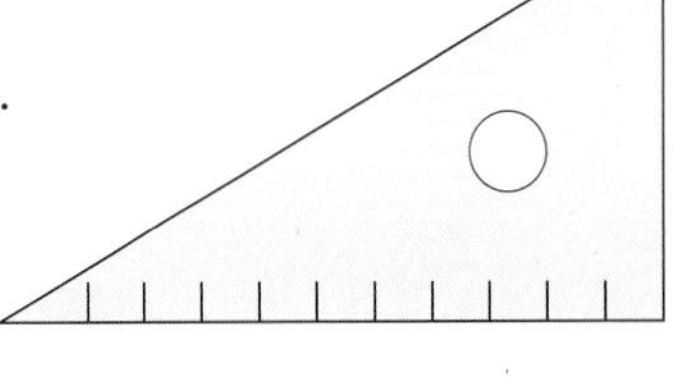

Can we make a square using some of these triangular rulers?

With the Way to See and Think Monsters...

Let's Reflect!

Let's reflect on which monster you used while learning " 2 Congruent Figures."

Why

We could explain why we can draw congruent quadrilaterals based on the properties of congruent figures.

① What are the properties of congruent figures?

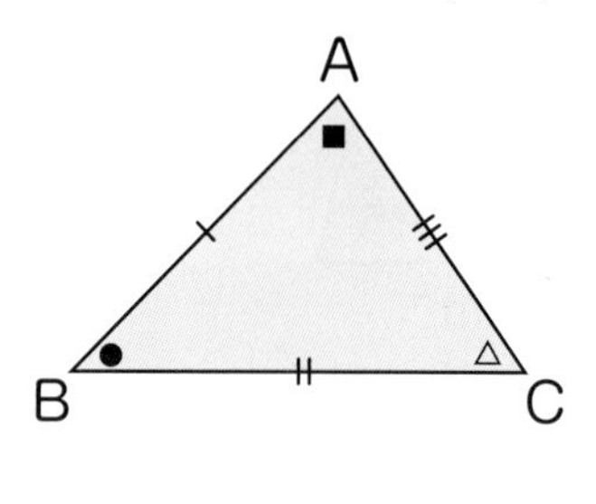

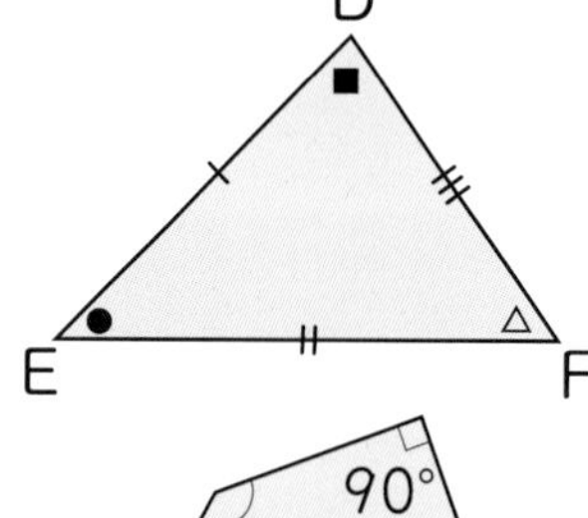

In congruent figures, the lengths of corresponding sides are equal and the size of corresponding angles are equal.

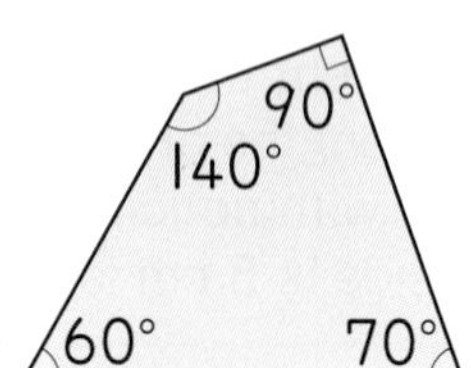

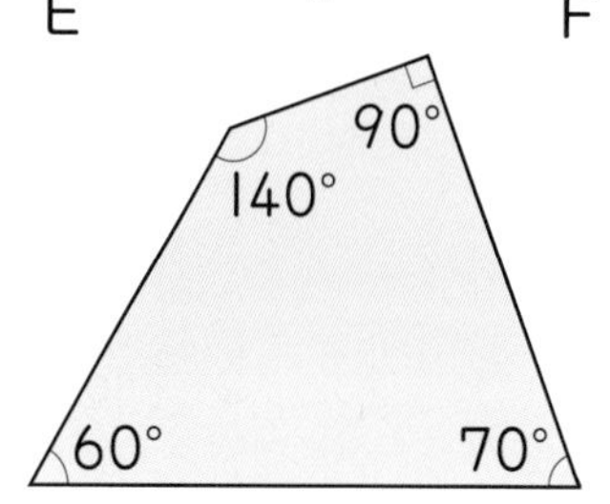

The two rectangles on the left have corresponding angles of equal size but the lengths of the corresponding sides are...

② How can you draw congruent quadrilaterals?

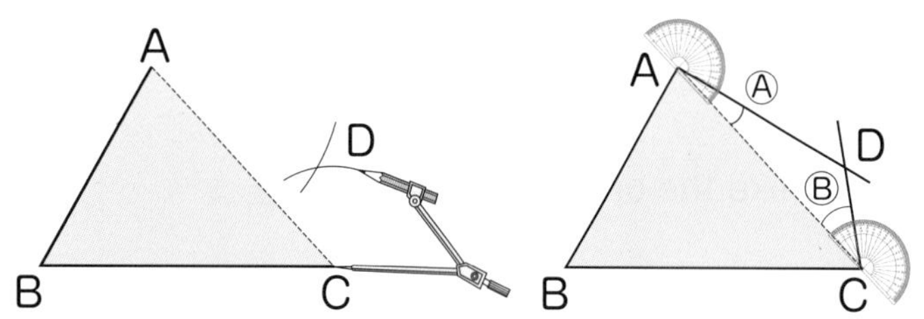

Drawing a diagonal line through a quadrilateral creates two triangles...

? Solve the ?

Using the properties of congruent figures, we could check whether two figures are congruent or not, and draw congruent figures.

Want to Connect

Can we find out the properties of other figures?

When explaining that figures are congruent and how to draw them, it is important to carefully explain the reasons in a step-by-step manner so that everyone can understand.
Be able to explain the reasons correctly.

Math Patrol

① In order to draw a triangle congruent to triangle ABC on the right, consider which side lengths and angle sizes of triangle ABC should be measured. The following Ⓐ~Ⓓ are marked with a circle where the length of a side or the size of an angle is measured. Choose one triangle that is congruent to triangle ABC.

A
B C

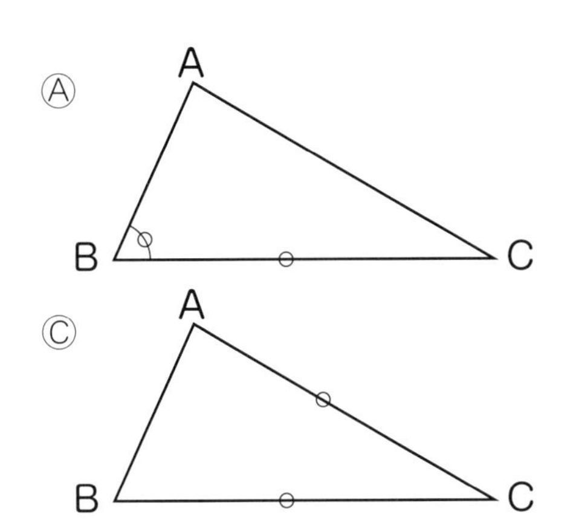

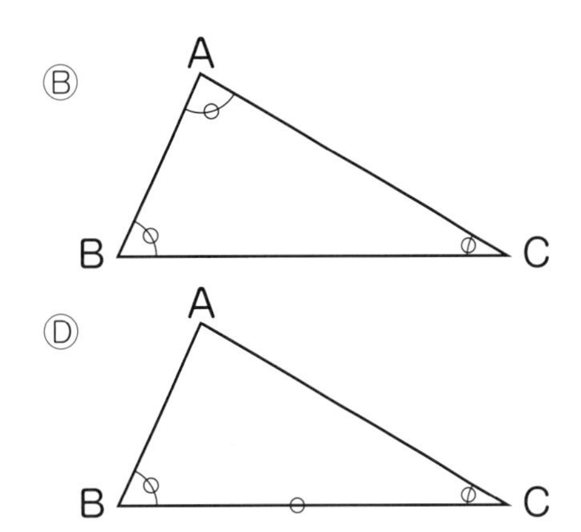

Frequently found mistake
Choose the option that the size cannot be set in one way.

Be careful!
If you can draw even one shape different from triangle ABC under the given conditions, it cannot be said that you can draw it. Let's check if there is any shape that is different from ABC can be drawn.

Try to explain why it is not possible to draw.

It cannot be said that congruent triangles can be drawn in Ⓐ since there are a number of possible positions for A.

A?
A?
A?
B C

In Ⓑ, we don't know the length of the sides...

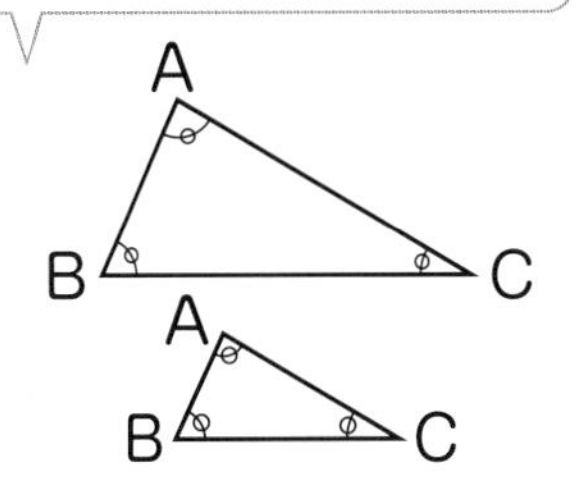

What are the rules for two quantities changing together?

\ Want to know /

What are the rules for the two quantities changing together?

Proportion

Let's explore quantities changing together and their correspondence.

1 Quantities changing together

Let's explore the following Ⓐ and Ⓑ.

Ⓐ What is the length and width of a rectangle with a perimeter of 30 m?

Ⓑ How many bricks are piled up and what is the total height when each brick has a height of 6 cm.

① Let's summarize the relationships of Ⓐ and Ⓑ in the following tables.

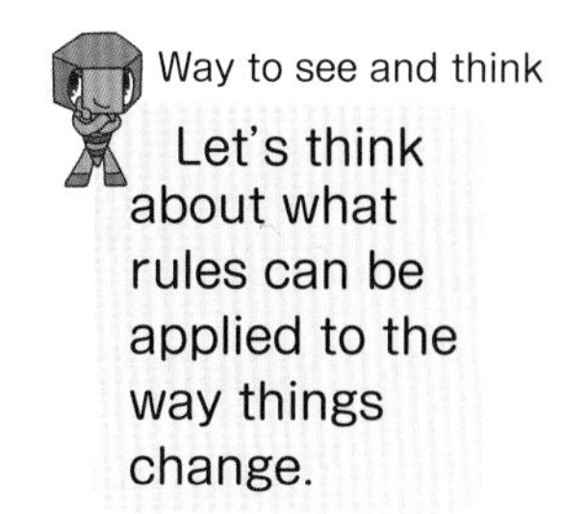

Way to see and think
Let's think about what rules can be applied to the way things change.

Length and width of the rectangle

Ⓐ

Length (m)	1	2	3	4	5	6	7
Width (m)	14	13					

Number of bricks piled up and height

Ⓑ

Bricks piled up	1	2	3	4	5	6	7	8
Height (cm)	6	12						

The way it changes is different.

When some quantities increase, some decrease.

② In Ⓐ, if the length increases, how does the width change?

③ In Ⓑ, if the number of bricks piled up increases, how does the height change?

Summary

Two quantities changing together have a relationship that increases as one quantity increases, and a relationship that decreases as one quantity increases.

? Is there any rule for how the two quantities change together?

1

There is a ribbon that costs 90 yen per meter. Let's explore the relationship between the length and the cost of the ribbon.

0 | 1 | 2 | 3 | 4 (m)

90 yen

90 yen 90 yen

90 yen 90 yen 90 yen

① Let's summarize the relationship between the length and the cost of the ribbon in the following table.

Length and cost of the ribbon

Ribbon's length (m)	1	2	3				
Ribbon's cost (yen)	90						

② How does the cost change as the ribbon's length increases 1m, 2m, ... and so on?

\ Want to explore /

? **Purpose** **What is the relationship between the ribbon's length and the cost?**

When the ribbon's length is □ m and the cost is ○ yen, if □ increases then ○ increases accordingly.

③ Let's explore how the corresponding cost ○ yen changes when the ribbon's length □ m increases 2 times, 3 times, 4 times, ..., and so on. Let's fill in each □ with the appropriate number.

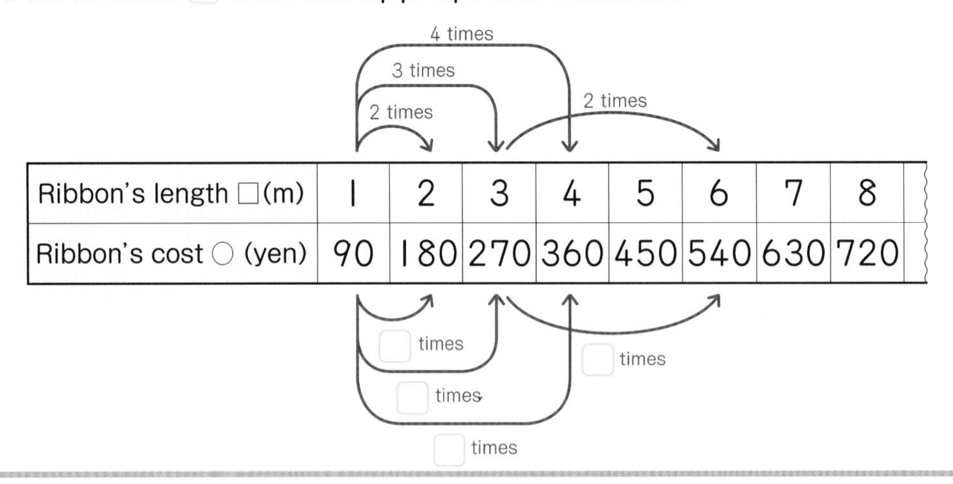

Ribbon's length □ (m)	1	2	3	4	5	6	7	8
Ribbon's cost ○ (yen)	90	180	270	360	450	540	630	720

If there are two quantities □ and ○ changing together, and □ **changes 2 times, 3 times**, ..., so on, and ○ **also changes 2 times, 3 times, ..., so on,** respectively. Then ○ **is called proportional to** □ .

Summary

If the ribbon's length changes 2 times, 3 times, ..., so on, and the cost also changes 2 times, 3 times, ..., so on. Then the cost is proportional to the ribbon's length.

4 How much is the cost when the length of the ribbon is 8 m?

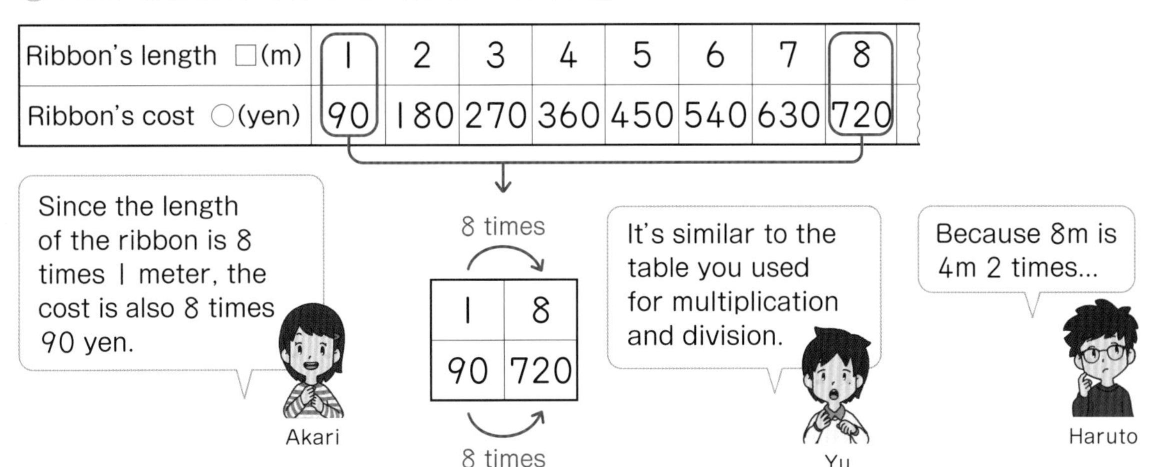

Ribbon's length □(m)	1	2	3	4	5	6	7	8
Ribbon's cost ○(yen)	90	180	270	360	450	540	630	720

Since the length of the ribbon is 8 times 1 meter, the cost is also 8 times 90 yen.
Akari
8 times
1 8
90 720
8 times
It's similar to the table you used for multiplication and division.
Yu
Because 8m is 4m 2 times...
Haruto

5 How much is the cost when the length of the ribbon is 12 m?

Also, what is the length of the ribbon when the cost is 1800 yen?

Sara
12m could be considered three times as long as 4m or twice as long as 6m.

Since we know the cost at 1m, can we formulate a multiplication equation using doubling?
Akari

1 Pile up bricks with a height of 6 cm. Let ○ cm be the total height after you pile up □ bricks. Let's answer the following questions.

① Let's summarize the relationship between number of bricks piled up and total height in the table.

Number of bricks piled up and height

Bricks piled up □	1	2	3	4	5	6	7	8
Total height ○(cm)	6							

Can you represent the relationship between □ and ○ in a math equation?

Yu

② The total height of the bricks is proportional to what?

③ What is the total height when 15 bricks are piled up? Also, how many bricks were piled up when the total height is 126 cm?

? Are there other proportional relationships?

2

As shown in the figure on the right, find the perimeter of a square that increases its side length by 1 cm. Let's think about the two quantities that change together.

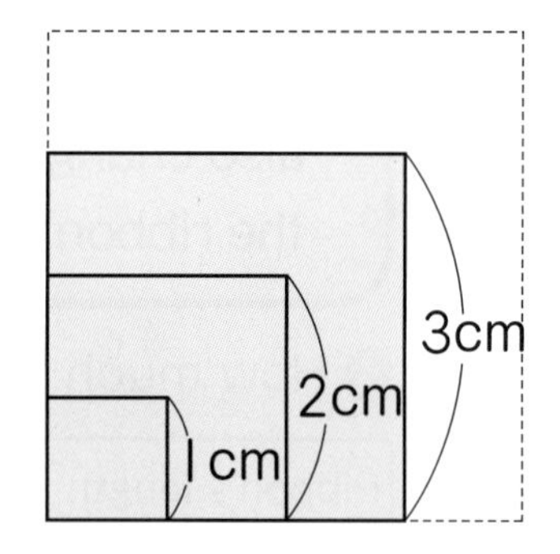

1. With word expressions, let's write a math equation for the perimeter, and explore which quantities change together.

2. What remains unchanged in the math equation 1 ?
3. Let's write a math equation for the perimeter, by using □ cm as the length of one side and ○ cm as the perimeter.
4. The following table summarizes the relationship between the length of one side □ cm and the perimeter ○ cm. How does ○ change when □ is increased 2 times, 3 times, 4 times, ..., so on? Let's fill in the □ with the appropriate number.

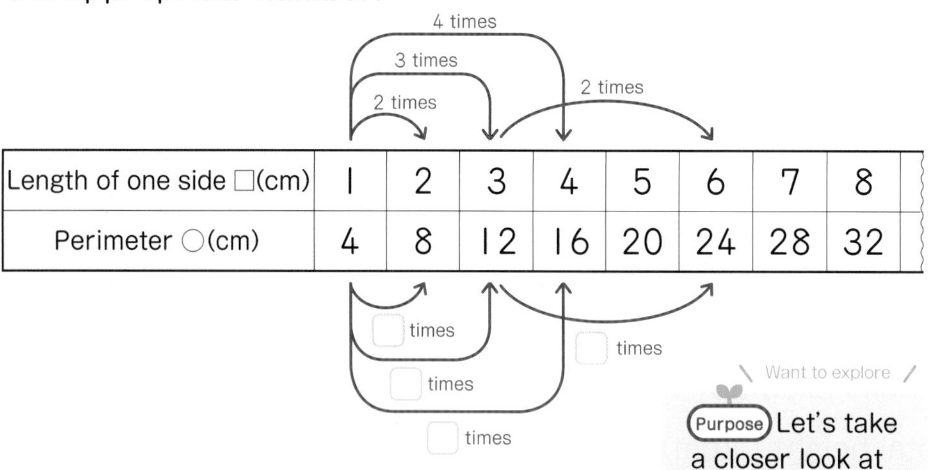

Length of one side □(cm)	1	2	3	4	5	6	7	8	
Perimeter ○(cm)	4	8	12	16	20	24	28	32	

5. Is the perimeter proportional to the length of one side? Let's also write the reasons.

Want to explore

Purpose Let's take a closer look at proportionality.

Haruto

6. If the perimeter of a square is 56 cm, then how many cm is the length of one side?

⑦ About the summary table in ④, we identify how ○ changes when □ is divided by 2, 3, ..., so on. Therefore, let's fill in the □ with the appropriate number.

÷4 ÷3 ÷2 ÷2

Length of one side □(cm)	1	2	3	4	5	6	7	8
Perimeter ○(cm)	4	8	12	16	20	24	28	32

÷□ ÷□ ÷□ ÷□

1 Rectangles with a 5 cm length and 3 cm width are connected as shown in the following diagram. Let's explore the relationship between the width and the area when the rectangles are connected.

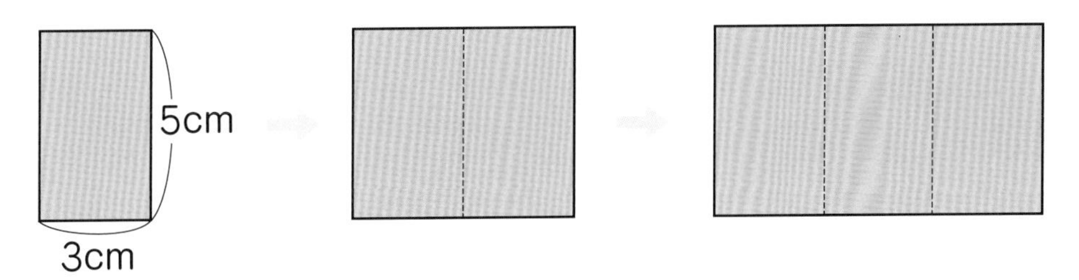

① Let's write a math equation for the area, use □cm as the width and ○ cm^2 as the area.

② Let's summarize the relationship between the width and the area of the rectangle in the following table.

Width and area of the rectangle

Width □ (cm)	3	6						
Area ○ (cm^2)								

③ Is the area of the rectangle proportional to the width? Let's also write the reasons.

Summary We don't have to write a entire table of proportional numbers to find the relationship, but we can find the proportions by using the double relation or by expressing them in a math equation.

Akari

CAN What can you do?

☐ We understand proportional relationships. → pp.38 ~ 39

1 In which case is ○ proportional to ☐ among the following Ⓐ to Ⓒ?

Ⓐ Length ☐ cm and width ○ cm when a rectangle's surrounding length is 26 cm.

Length ☐(cm)	1	2	3	4	5	6
Width ○(cm)	12	11	10	9	8	7

Ⓑ ☐ number of balls and the total cost ○ yen when the cost of one ball is 300 yen.

Number of balls ☐	1	2	3	4	5	6
Total cost ○(yen)	300	600	900	1200	1500	1800

Ⓒ ☐ number of candies and the total cost ○ yen when the cost of one candy is 8 yen.

Number of candies ☐	1	2	3	4	5	6
Total cost ○(yen)	8	16	24	32	40	48

☐ We can solve problems on proportion. → pp.40 ~ 41

2 Water is being poured into a water tank so that the depth of water increase 2 cm in 1 minute. Let's explore the relationship between the time ☐ (min) to pour water and the depth of water ○ cm.

① Let's explore the relationship between the time ☐ (min) to pour water and the depth of water ○ cm in the table.

Time to pour water and the depth of water

Time to pour water ☐(min)	1	2	3	4	5	6
Depth of water ○(cm)	2					

② Which is proportional to what?

③ If ☐ increases by 1, how much does ○ increase?

④ Let's represent the relationship of ☐ and ○ in a math equation.

⑤ Let's find out the depth of water when the time to pour water is 9 minutes.

⑥ Let's find out the time to pour water when the depth of water is 30 cm.

Supplementary Problems → p.158

Which "Way to See and Think Monsters" did you find in "3 Proportion"?

I found "Why" when I was exploring the relationship between the two changing quantities.

Yu

I found other monsters, too!

Akari

With the Way to See and Think Monsters...

Let's Reflect!

Let's reflect on which monster you used while learning "3 Proportion."

Why

The proportionality rule could be used to explain whether two quantities changing together can be said to be proportional or not.

① There is a ribbon that costs 90 yen per meter. When the ribbon's length is □ m and the cost is ○ yen, ○ is called proportional to □. Let's think about the reason why we can say so.

Ribbon's length □(m)	1	2	3	4	5	6	7	8
Ribbon's cost ○(yen)	90	180	270	360	450	540	630	720

Sara: The price for 1m ribbon is 90 yen, for 2 m ribbon is 180 yen, ... and when the length of ribbon is □ times, □ times, ..., the price will be □ times, □ times, ... too. The price is proportional to the length of the ribbon.

Yu: Using the rule of proportions, I can find the cost of a ribbon for any length.

② What else have you noticed while learning proportions?

Akari: It was as if the multiplication table had been taken out of some of the tables of proportions.

Haruto: When ○ is proportional to □, the formula to find ○ is always multiplication. Is there a rule?

? Solve the ?

Given two quantities □ and ○ that change with each other, we found out that there are times when ○ is proportional to □ and times when it is not.

Akari

Want to Connect

Besides proportionality, what is the relationship between the two quantities that change with each other?

Sara

How much juice can we make?

▷

We have a lot of oranges. So let's make orange juice!
オレンジ
I wonder how much juice can we make from 1 orange?
I poured the juice in 5 containers but each of them has a different amount.
Can we determine the amount of juice that can be made from one orange?
How much juice will be made if all of it is made into juice?

How can I find out how much juice can be squeezed from one orange?

Mean
Let's think about how to make each amount the same.

1 After squeezing 5 oranges, the result is shown in the figure on the right. Let's consider how much juice can be squeezed from one orange.

A	B	C	D	E
85mL	95mL	90mL	110mL	70mL

Why don't you transfer from too many to a few?

I would like to pour all juice in one big container and then divide it.

① Let's compare the ideas of the following children.

Akari's idea

Move juice from the one with more amount to the one with a less amount.

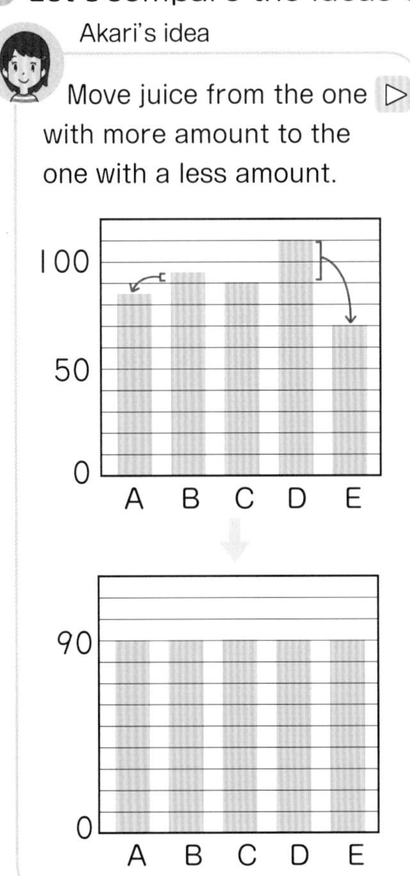

Yu's idea

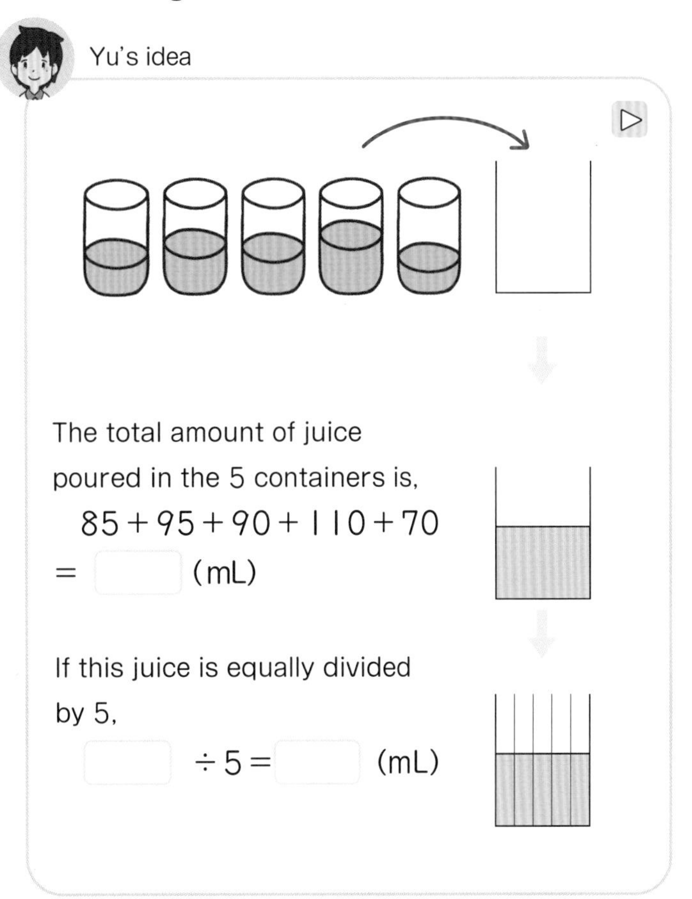

The total amount of juice poured in the 5 containers is,

$85+95+90+110+70$

$=$ ☐ (mL)

If this juice is equally divided by 5,

☐ $\div 5=$ ☐ (mL)

The same number or measure which was obtained by getting the average of the given numbers or measures is called the **mean**.

Mean = Total ÷ Number of items

If we try to represent Yu's thinking method, it becomes as follows:

$(85+95+90+110+70) \div 5 = 90$

Total amount of juice — Number of containers — Mean

By adding several values and dividing them equally by the number of pieces to obtain the mean, we can find out the value of each piece.

② If there are 25 oranges in total, using the average obtained above, how much juice can be squeezed out of all of them?

1 The following table shows the number of books that Soshi read from April to August. What is the mean number of books he read in one month?

Number of books read

Month	April	May	June	July	August
Number of books	4	3	0	2	5

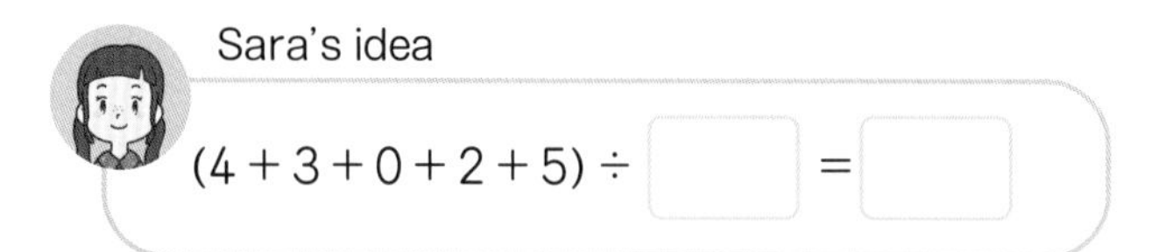

He didn't read any book in June, so I wonder if this affects the total number of months.

Words

【平】 Looking flat and even.

【均】 Making things equal.

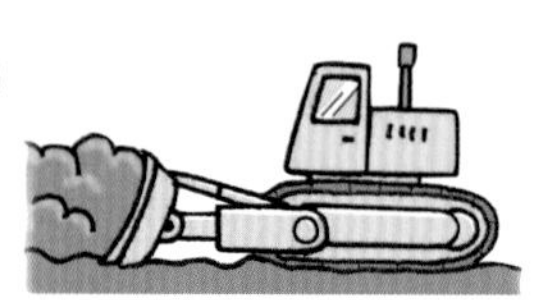

Even if something can not be expressed in a decimal number, such as numbers of books, the mean might be expressed in a decimal number.

? When finding out the mean, what should we do when the total number of items is different?

2 A group of 5th graders run in the playground every morning. The following table summarizes how many laps Erika,Yuji, and Nozomi ran around the playground last week. Nozomi rested for one day, so she only ran 4 days. Who ran more laps around the playground?

Nozomi was the only one who did not run 1 day.

Sara

Erika's number of laps

Day number	Day 1	Day 2	Day 3	Day 4	Day 5
Number of laps	1	3	4	4	3

Yuji's number of laps

Day number	Day 1	Day 2	Day 3	Day 4	Day 5
Number of laps	5	3	7	2	3

Nozomi's number of laps

Day number	Day 1	Day 2	Day 3	Day 4
Number of laps	7	4	5	4

Want to know

? Purpose What should we do if the number of days are different?

① Let's look at the records of the 3 children on the previous page and discuss.

If we try to compare in total...
Akari

But Nozomi's total days is 4. Is it good to compare in total even though the number of days is smaller?
Haruto

If Nozomi had not taken a rest, how many times would she have run that day?
Yu

② As for the 3 children, what is the mean of times they ran in one day?

1 The following table shows the scores of the 1st and 2nd group of students that took a math test. Answer the questions below.

Scores of the 1st group

Name	Yuri	Kota	Saki	Keiya	Rinka
Score (points)	78	65	70	81	90

Scores of the 2nd group

Name	Sho	Tomomi	Masaya	Chiaki	Itsuki	Fumino
Score (points)	82	63	69	74	88	86

The mean score is the average number of points scored per person.

① Let's find the mean scores of the first and second group of students in the math test.

② Which group has a higher mean score?

Summary

Even if the number of days or the number of people are different, we can find the mean of each and compare them.

? When is the mean useful?

Words

【 If ..., then ... is .... 】

This is a word that you use when something is assumed or predicted. In mathematics, it is often used when the conditions are altered to get the conclusion.

3 Let's find out the approximate length of one step.

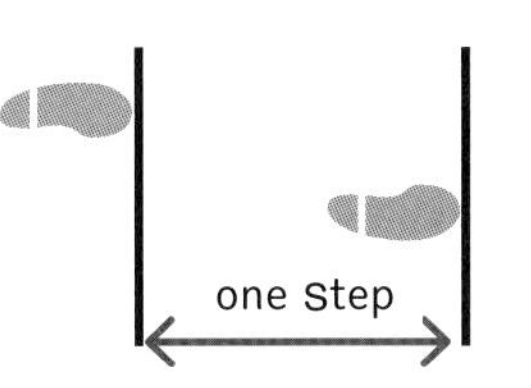

\ Want to know /

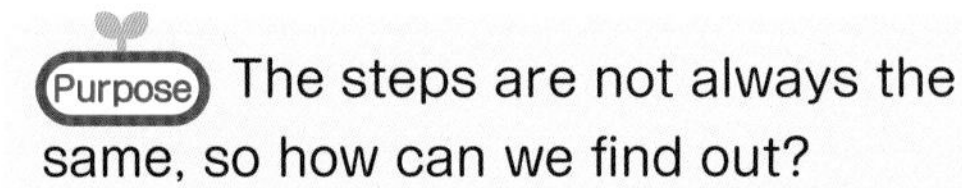

Purpose The steps are not always the same, so how can we find out?

Haruto

① The following table shows the distance Kasumi walked with 10 steps. How many meters is one step? Let's find it out by rounding to the nearest hundredth.

Distance Kasumi walked in 10 steps

Number of times	1st time	2nd time	3rd time
Distance walked with 10 steps (m)	5.17	5.13	5.18

Mean number of the 3 times:

$(5.17 + 5.13 + 5.18) \div 3 = \square$

One step $5.16 \div 10 = \square$ Approximately $\square$ m

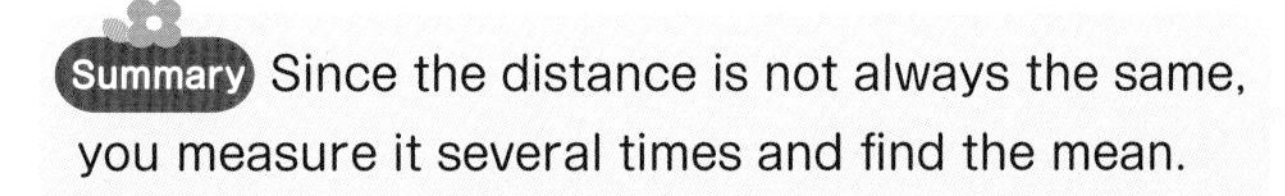

Summary Since the distance is not always the same, you measure it several times and find the mean.

Akari

② Kasumi counted 1731 steps from home to school. Using the result from ①, can you think how many meters is the approximate distance from home to school? Let's find it out by rounding to the nearest whole number.

③ Let's explore the length of your own step, and try to explore the approximate distance of different places.

What is the distance in meters if you go around the school?

Yu

? Are there other situations when we can use the mean?

During a science experiment, Megumi examined the time it takes a pendulum to go back and forth (period) 10 times. Let's find the mean for 1 period of the pendulum.

Times it takes to go back and forth

Number of times	1st time	2nd time	3rd time	4th time	5th time
Time for 10 periods (sec)	15	23	14	13	14
Time for 1 period (sec)	1.5	2.3	1.4	1.3	1.4

① Let's look at the above table, and discuss how to find the answer.

Haruto: There is a value far from the other values.

Akari: I wonder if we can use that value the same way...

\ Want to think /

? **Purpose** How should we deal with values that are far apart?

② Yu's idea is as follows. Let's explain his idea.

Yu's idea

Since I think the 2nd time failed. The mean should be found by excluding this value. So, for this experiment, the mean should be found using the 1st, 3rd, 4th and 5th time.

Therefore, divided in ☐ parts.

(☐ + ☐ + ☐ + ☐) ÷ ☐ = ☐

First of all, let's explore why the outlier came out.

If you find the mean with all the values, (1.5+2.3+1.4+1.3+1.4) ÷ 5=1.58, then it is about 1.6 seconds.

Sara

! **Summary**

When there is a large distant value, the result may be affected, so in that case it is better to calculate the mean by excluding that value.

1. The following table shows the records when Arata ran 50 m. Using the mean, think about how many seconds Arata can run in approximately 50 m.

Arata's record for a 50 m run

Number of times	1st time	2nd time	3rd time	4th time	5th time
Record (sec)	9.5	9.1	9.4	12.6	9.2

The 4th time record is an outlier.

Haruto

? When the number of items is big, it is likely that you will make a mistake in your calculation. I wonder if we can improve the method...

5 Consider chicken Ⓐ and Ⓑ. Which chicken laid heavier eggs?

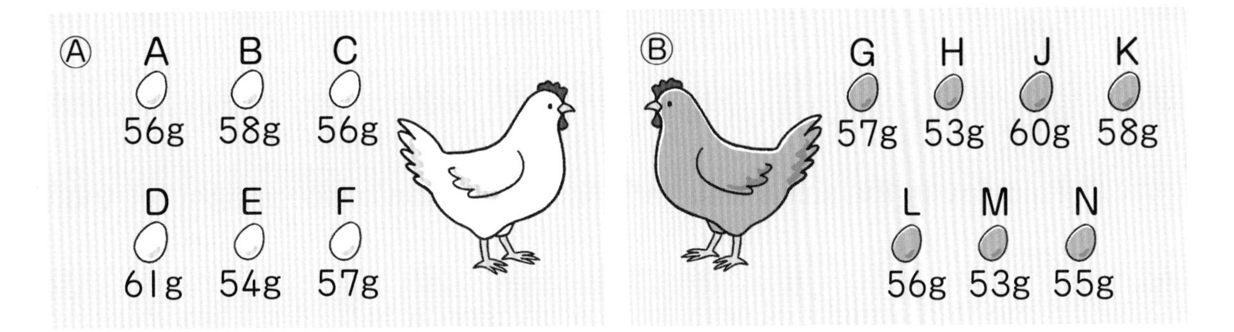

1. Let's find and compare the mean of the weights.

Yu: Both chickens are laying a lot of eggs, aren't they?

Sara: The lightest egg of chicken Ⓐ weighs 54g, and the heaviest weighs 61g.

\ Want to think /

? **Even when the number of items is large or total is large, can we find the mean without making a mistake?**

② Sara came up with the following idea to find the mean of the weight of the eggs of chicken Ⓐ. What was the improvement made by Sara?

Sara's idea

Taking as reference the lightest egg, 54 g, I thought about how much heavier the other eggs were.

56	58	56	61	54	57 (g)
↓	↓	↓	↓	↓	↓
2	4	2	7	0 (Reference)	3 (g)

I found the mean by,

$(2+4+2+7+0+3) \div 6 = 3$

If the mean is added to the weight I used as a reference, I can find the mean of the weight of the eggs.

$54 + 3 = 57$

Then, the mean of the weight of the eggs laid by Ⓐ is 57 g.

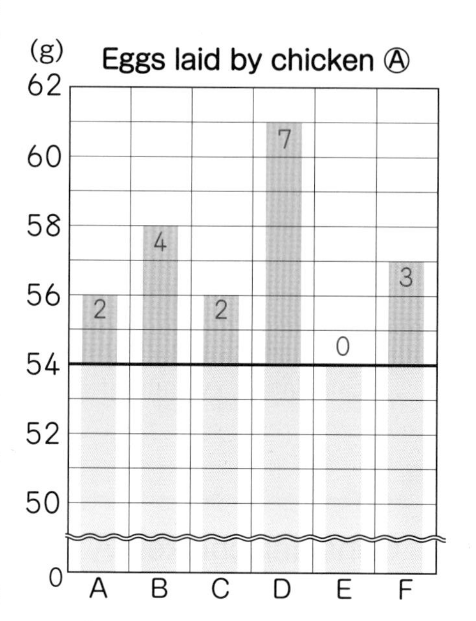

You are thinking about how much heavier it is compared to the lightest of 54 g.

Even for the things that you can't distribute evenly, if you know the number or the amount, you can calculate the mean. The value used as a reference, such as 54 g, is called the "assumed mean."

③ Thinking in the same way, let's find the mean of the weight for the eggs of chicken Ⓑ.

④ If the total weight of eggs laid by chicken Ⓑ is 728 g. Can you calculate how many eggs were laid?

Summary

Using an assumed mean can make the process of finding the mean easier.

1 The table on the right shows the weight of 6 potatoes that I bought at the supermarket. Let's find the mean weight of one potato.

Weight of potatoes

157 g, 155 g, 143 g,
152 g, 148 g, 145 g

CAN What can you do?

☐ We can find out the mean. → p.45 ~ 48

1 The following table shows the number of empty cans picked by Marina in 5 days.

Number of empty cans picked

Day number	1st day	2nd day	3rd day	4th day	5th day
Number of empty cans	6	7	5	9	8

① What is the mean of the number of cans picked in one day?

② If Marina picks up empty cans in the same way for 15 days, can she calculate the total number of empty cans picked?

☐ We can use the mean. → p.49

2 Toshiki walked 1850 steps from his house to the library. If Toshiki's one step is about 0.54 m, can you calculate the distance from Toshiki's house to the library in meters?

☐ We can find out the mean when measuring. → pp.50 ~ 51

3 The table on the right shows the records for Sayaka's softball throws.
Using the mean, consider how far Sayaka can throw a softball in meters (m).

Record for softball throws

Number of times	1st time	2nd time	3rd time	4th time
Records(m)	24	22	12	26

☐ We can solve problems using the mean. → pp.47 ~ 48

4 The 1st and 2nd group of 5th grade went to a potato field. The number of people in each group and the total number of potatoes harvested are shown in the table on the right.
Can you say which group harvested the most number of potatoes? What is the mean number of potatoes harvested by 1 person? Let's compare.

Number of potatoes harvested

	Number of people	Total number
1st group	24	84
2nd group	30	102

Supplementary Problems → p. 158

Which "Way to See and Think Monsters" did you find in "4 Mean"?

I found "Align" when I was thinking about the mean.

Akari

I found other monsters, too!

Sara

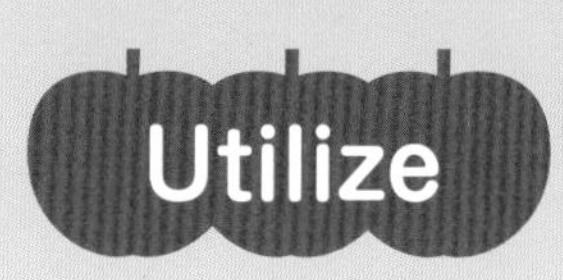

Usefulness and Efficiency of Learning

1 The following table shows the number of tomatoes harvested by pot planing from Monday to Friday. What is the mean number of tomatoes harvested in one day?

Number of tomatoes picked from potted plants

Day of week	Monday	Tuesday	Wednesday	Thursday	Friday
Number of tomatoes	6	3	2	0	8

2 Kazuya's aim is to read the mean number of 25 pages a day. The mean number for 6 days (from Sunday to Friday) was 23 pages. How many pages should Kazuya read on Saturday, so that in a 7 days period (from Sunday to Saturday), he achieves his objective to read rhe mean number of 25 pages a day?

3 The following table shows the records for Yuki's long jumps.

Records for long jumps

Number of times	1st time	2nd time	3rd time	4th time	5th time	6th time
Record	2m45cm	2m32cm	2m44cm	2m48cm	1m25cm	2m36cm

Using the mean, let's think about how far Yuki can jump in m and cm.

4 There are 6 onions inside a box. The weights can be seen on the box on the right. Let's fill in each □ with the proper answer.

162 g, 157 g, 159 g,
161 g, 160 g, 158 g

The mean weight of the onions inside the box was found as follows. Taking as reference the smallest weight, 157 g, think about how much heavier the other onions are. We found the mean by,

162	157	159	161	160	158 (g)
↓	↓	↓	↓	↓	↓
□	□	□	□	□	□ (g)

(□ + □ + □ + □ + □ + □) ÷ 6 = □

so the mean weight of the onions was,

157 + □ = □ □ g

With the Way to See and Think Monsters...

Let's Reflect!

Let's reflect on which monster you used while learning "4 Mean."

Align

By adjusting the pieces so that they are equally distributed, we were able to find the average per piece.

① To find out how much juice can be squeezed from 1 orange, we used the juice of 5 oranges. Then, how did you figure out how much juice could be squeezed from one orange?

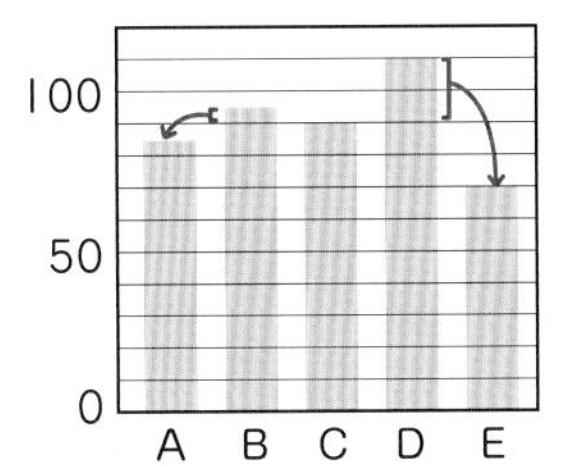

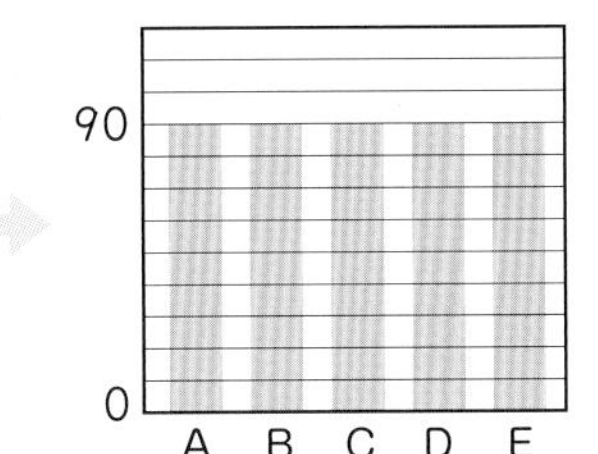

We transferred from the larger amount to the smaller one until all of them were equally distributed.

Akari

Summarize

Divide

By adding the whole quantity and then dividing it by the number of items, we were able to obtain the mean.

After putting the five juices together, we divided them into five equal parts and came up with one piece.

Yu

Let's deepen. → p.167

Solve the ?

By finding the mean, we were able to find approximately how much one piece would be.

Sara

Want to Connect

Are there other situations in which we can think in terms of mean?

Haruto

Find the ? Let's divide into teams.

Let's divide into two teams, red and white, to play dodgeball.
There are 32 students in the class.
How should we divide?
Since we just need to divide in half, 32 ÷ 2 = 16, each team should have 16 students.
Then, I wonder if they should be divided into 1～16 and 17～32 by student number.
There is also another way to divide. First person is red, second person is white, third person is red, ..., and so on. You can divide them in order.

\ Want to know /

Purpose When categorizing numbers in order, is there any rule?

Multiples and Divisors

Let's think about the multiples and divisors.

1 Even and odd numbers

How to categorize whole numbers →

1 Divide 32 children in the class into red and white teams. When you divide "red, white, red, white, ..." by the student number, can you identify what kind of number are grouped in each team?

① What kind of numbers are the red and white teams grouped?
Let's write the student number and complete the following table.

Red	1	3	5
White	2	4	6

Akari: Red and white number are increasing by 2.

Yu: I wonder what kind of classification is it?

② Let's try to divide the student number by 2 for the red and white team.

Red team

$1 \div 2 = 0$ remainder 1
$3 \div 2 = 1$ remainder 1
$5 \div 2 = 2$ remainder 1
⋮

White team

$2 \div 2 = 1$
$4 \div 2 = 2$
$6 \div 2 = 3$
⋮

Way to see and think
We can divide it into group with remainder 1 and group with remainder 0.

③ What is the team for children with student number 11 and 18?

④ Let's think what kind of classification rules are there for each team.

For whole numbers, the numbers that are divisible by 2 are called **even numbers**. While the numbers that are not divisible by 2 are called **odd numbers**. 0 is an even number.

Even and odd numbers have the following representation.

Even number: number multiplied by 2. $2 \times \square$

Odd number: Add 1 to a number multiplied by 2. $2 \times \square + 1$

We can see from the equation that odd numbers have remainder 1 when is divided by 2.

Yu

1 How are even and odd number aligned? Let's explore, write a ○ on even numbers, and □ on odd numbers in the following number line. Is there any whole number that is not classified as even or odd numbers?

Whole Number

Even numbers	Odd numbers
0, 2, 4, 6, …	1, 3, 5, 7, …

Way to see and think

All the whole numbers can be classified as even or odd number.

2 Let's classify the following whole numbers as even or odd numbers. Also, explain how to tell them a part.

98 99 100 217 218 1234

3 From your surroundings, let's find occasions where even and odd numbers are used.

航空会社 Airline	便名 Flight No.	出発地 Origin
AN	840	中標津
エア	20	札幌
AN	4720	札幌
AN	256	福岡
AN	572	稚内
AN	678	広島
AN	536	高松
AN	876	秋田

行先	便名	天気	搭乗
大阪 / 伊丹	125		11
札幌	533		18
大阪 / 関西	1011		21
札幌	4643		16
大阪 / 伊丹	4631		15
小松	1281		17
札幌	537		14
三沢	1229		34
大阪 / 伊丹	131		16

The number of flights to Tokyo is an even number, and the number of flights from Tokyo is an odd number.

? Are there any other rules for even and odd numbers?

2

Even and odd numbers are represented on the diagram on the right. Answer the following questions.

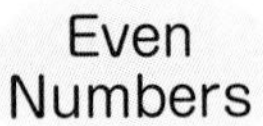
Even
Numbers

Odd
Numbers

2
4
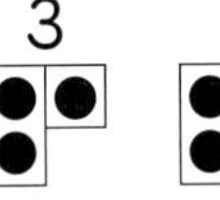
3
5

① Let's represent some of the odd numbers in a diagram like the one above. Let's discuss our findings.

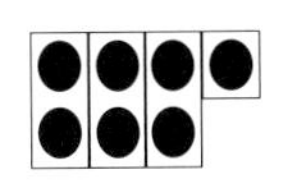
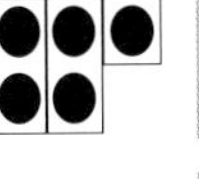

7 is an odd number.
The diagram on the right shows that 7 as an odd number.
Let' s try to compare the diagram for 3 and 5...
Haruto

② Looking at the diagram above, Haruto thought that the sum of odd numbers is an even number. So, let's find the sum of several odd numbers and verify it.

Yu
3＋5＝8,
9＋13＝22, ...

I wonder if it would also be an even number with bigger odd numbers...
Akari

Want to explain
Purpose

Can you explain how the sum of odd numbers can always be an even number?

③ Haruto came with the following idea to explain why the sum of odd numbers is an even number. Let's explain his idea.

Haruto's idea

If we make a diagram for the odd numbers, we can always a group in pairs with the ones that are left. In other words, if you add odd numbers, you will always have 1 left, and because there are two 1 left, and this makes 2.

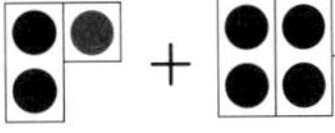
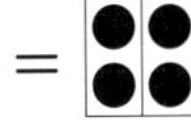
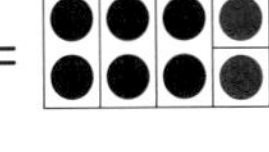
+
=

Summary

Using the diagram, we can illustrate that the sum of odd numbers is always an even number.

④ Let's discuss what kind of numbers we get when we add even and odd and when we add two even numbers.

Numbers that are divisible by 2 were called even, but I wonder if numbers divisible by 3 or 4 have names?

【Rules of the Clap Number Game】

Let's form a circle. Decide the person that will be the first one. Say out loud the numbers in order starting from 1, and clap your hands on each set of the "Clap Number". For example, when the "Clap Number" is 3, then every 3 numbers, clap the hands while saying the corresponding number out loud.

Multiples ↓

1

Group A decided the "Clap Number" is 3. Let's think about which numbers will have clapping hands.

① Let's write the numbers in the table shown on the right, and color the numbers that will have clapping hands.

1	2	3	4	5	6	7	8	9	10
11	12	13	14	15	16	17	18	19	20
21	22								

② Check the colored numbers, and let's identify what kind of numbers are grouped.

The colored numbers are 3, 6, ...

Haruto

The numbers are the same as in the multiplication table of 3.

Sara

\ Want to explore /

? **Purpose** **What characteristics do you see in the numbers when you clap the hands?**

Ⓐ Multiples of 2 — 0 1 2 3 4 5 6 7 8 9 10 11 12 13 14 15 16 17 18 19 20 21 22 23 24 25 26 27

Ⓑ Multiples of 3 — 0 1 2 3 4 5 6 7 8 9 10 11 12 13 14 15 16 17 18 19 20 21 22 23 24 25 26 27

Ⓒ Multiples of 4 — 0 1 2 3 4 5 6 7 8 9 10 11 12 13 14 15 16 17 18 19 20 21 22 23 24 25 26 27

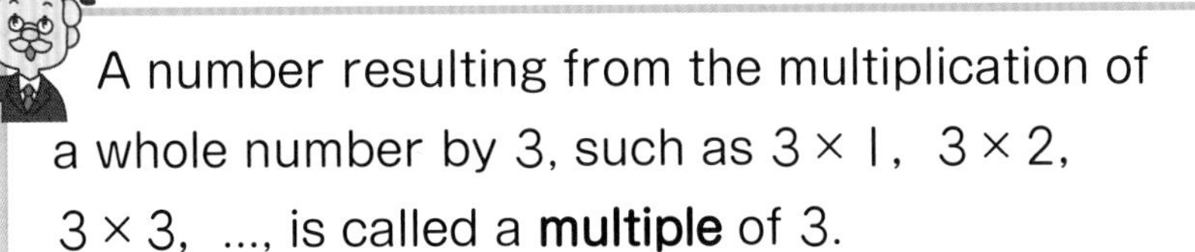

A number resulting from the multiplication of a whole number by 3, such as 3×1, 3×2, 3×3, ..., is called a **multiple** of 3.
0 of 0×3 is not a multiple of 3.

Multiples of 3
3 6 9 12
15 18 …

③ Let's place a ◯ on the multiples of 3 in the number line Ⓑ at the bottom of this page.

Summary

If the "Clap Number" is 3, then every time you clap your hands is a multiple of 3.

1 Let's place a ◯ on the multiples of 2 and 4 in the respective number lines Ⓐ and Ⓒ at the bottom of this page.

2 Let's answer the following questions when stacking boxes with a height of 5 cm.

① What is the total height in cm when 6 boxes are stacked?

② The total height is multiple of what number?

3 Let's find the smallest 5 numbers that are multiples of the following.

① multiples of 7 ② multiples of 8

③ multiples of 9

Can you express this in a multiplication equation?
Akari

4 The following numbers are multiples of what number?

① 10 ② 12 ③ 24

? In the number line below, what numbers are multiple of 2 and multiple of 3 at the same time?

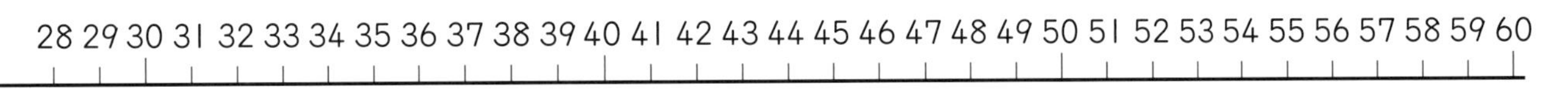

28 29 30 31 32 33 34 35 36 37 38 39 40 41 42 43 44 45 46 47 48 49 50 51 52 53 54 55 56 57 58 59 60

28 29 30 31 32 33 34 35 36 37 38 39 40 41 42 43 44 45 46 47 48 49 50 51 52 53 54 55 56 57 58 59 60

2

In the Clap Number Game, Group A clap hands when it is a multiple of 3, and Group B clap hands when it is a multiple of 4. Let's discuss what you noticed when you play the game.

Group A clap hands as follows:

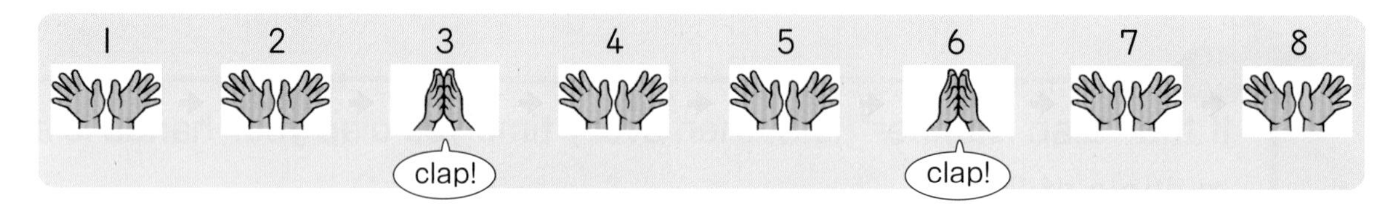

Group B clap hands as follows:

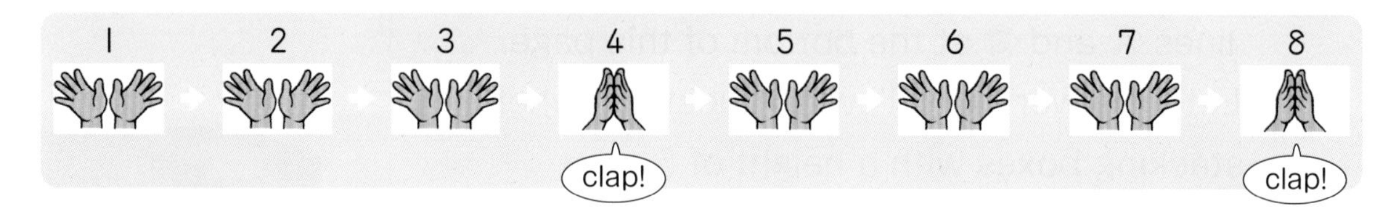

3, 6, 9, ... are multiples of 3.

Is there a number in which Group A and B clap hands at the same time?

\ Want to explore /

? Purpose On which number do they clap hands at the same time?

① In the following table, let's write ○ for the number in which Group A and B clap hands. And let's write × for the number in which they do not clap hands.

	1	2	3	4	5	6	7	8	9	10	11	12
Group A	×	×	○	×								
Group B	×	×	×	○								

② When do both groups clap hands at the same time?

③ If they continue the game, when is the second time that both groups clap hands at the same time?

A number that is a multiple of both 3 and 4 is called a **common multiple** of 3 and 4. The smallest of all common multiples is called **the least common multiple**.

④ Let's find 5 common multiples of 3 and 4, in ascending order. Also, what is the least common multiple of 3 and 4?

You can use the number lines from page 60 and 61.

Multiples of 3 3, 6, 9, 12, 15, 18, 21, 24, …

Multiples of 4 4, 8, 12, 16, 20, 24, 28, …

Least common multiple   Common multiple

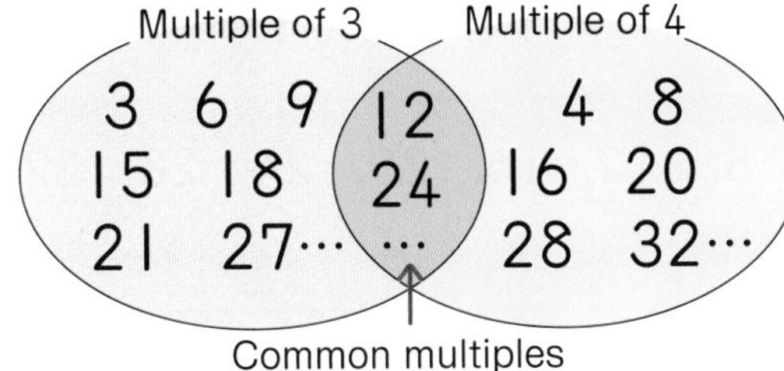

Way to see and think

Similarly, as there are may multiples of a number, there are also many common multiples.

Summary

If you clap hands in multiples of 3 and 4, then they will clap hands at the same time in 12, 24, .. etc., are the common multiples of 3 and 4.

? I wonder if we can figure out a strategy to find the common multiples?

That's it! Make tapes that find common multiples

When a tape with holes on the multiples of 2 and a tape with holes on the multiples of 3 are pilled up, the number of the places were they overlap are common multiples of 2 and 3. Let's make tapes with holes on more multiples to find more common multiples.

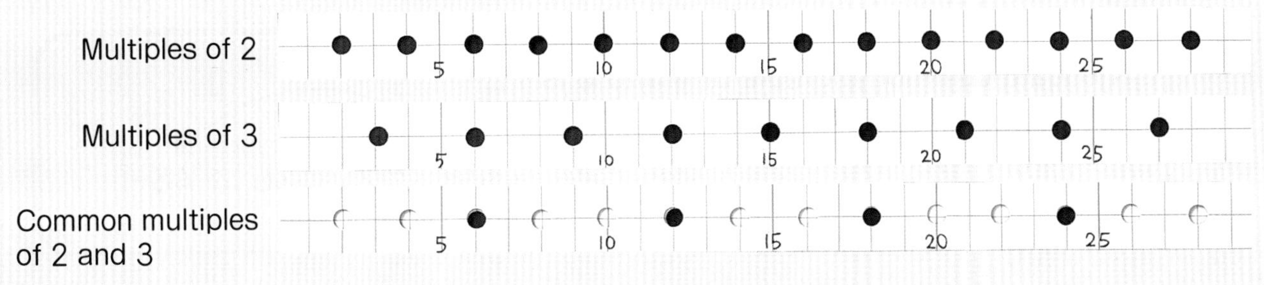

Let's think about how to find the common multiples of 4 and 6.

＼ Want to think ／

? Purpose How can we find common multiples?

① Let's explain the way of thinking of the following 3 children.

Yu's idea

Multiples of 4　4, 8, (12), 16, 20, (24), 28, 32, (36), 40, …

Multiples of 6　6, (12), 18, (24), 30, (36), 42, 48, 54, 60, …

Akari's idea

Multiples of 4	4,	8,	12,	16,	20,	24,	28,	32,	36, …
	×	×	○	×	×	○	×	×	○

Sara's idea

Multiples of 6	6,	12,	18,	24,	30,	36,	42,	48, …
	×	○	×	○	×	○	×	○

② What is the least common multiples of 4 and 6?

③ Haruto looked at the common multiples of 4 and 6 and the least common multiple and thought as follows. Explain his idea. Also, check if the numbers he found are common multiples or not.

Haruto's idea

4, 8, (12)
6, (12)　　$12\times2=(24)$　　$12\times3=(36)$　　$12\times4=(48)$　　$12\times5=(60)$

! Summary

The least common multiple of 4 and 6 is 12.

The common multiples of 4 and 6 are also multiples of the least common multiple.

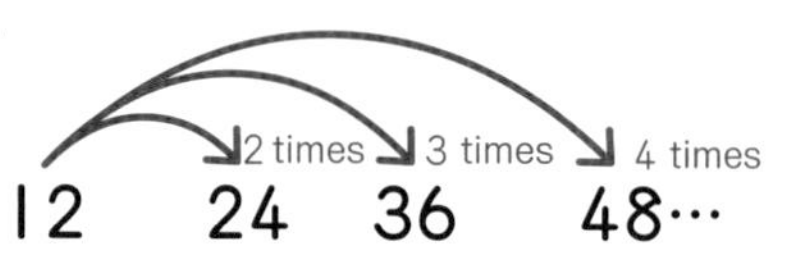

1 Let's find the first 4 common multiples of the following set of numbers.

① (5, 2)　② (3, 9)　③ (6, 9)

2 Yu came up with the following idea about the least common multiple and the common multiples of 2 and 6. Is this idea correct? If not, let's write down the reasons.

Yu's idea

The least common multiple of 3 and 4 is 12. This can be found by 3 x 4=12. Therefore, since 2 x 6=12, the least common multiple of 2 and 6 is also 12. Thus, the common multiples of 2 and 6 are multiples of 12.

? I wonder if we can also find the common multiple of 3 numbers?

Common multiple of 3 numbers ↓

4 Let's find the common multiples of 2, 3, and 4.

\ Want to think /

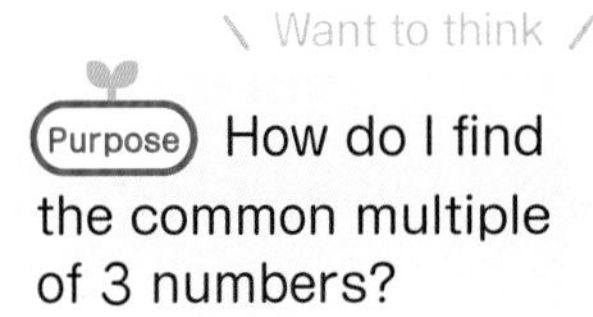

Purpose How do I find the common multiple of 3 numbers?

Sara

① Let's place a ○ on the multiples of 2, 3, and 4 in each of the following number lines.

Multiples of 2　0 1 2 3 4 5 6 7 8 9 10 11 12 13 14 15 16 17 18 19 20 21 22 23 24

Multiples of 3　0 1 2 3 4 5 6 7 8 9 10 11 12 13 14 15 16 17 18 19 20 21 22 23 24

Multiples of 4　0 1 2 3 4 5 6 7 8 9 10 11 12 13 14 15 16 17 18 19 20 21 22 23 24

② What is the least common multiple of 2, 3, and 4?

③ Let's find the first 3 common multiples of 2,3, and 4.

Summary We can find the common multiple of 3 numbers using the same method used to find the common multiple of two numbers.

Akari

1 Let's find the first 4 common multiples of the following set of numbers.

① (3, 5, 6)　② (4, 7, 14)　③ (5, 10, 15)

2 Let's make a square by aligning 5 cm long and 6 cm wide rectangular paper in the same direction as shown on the right.
Let's answer the following questions about the resulting square.

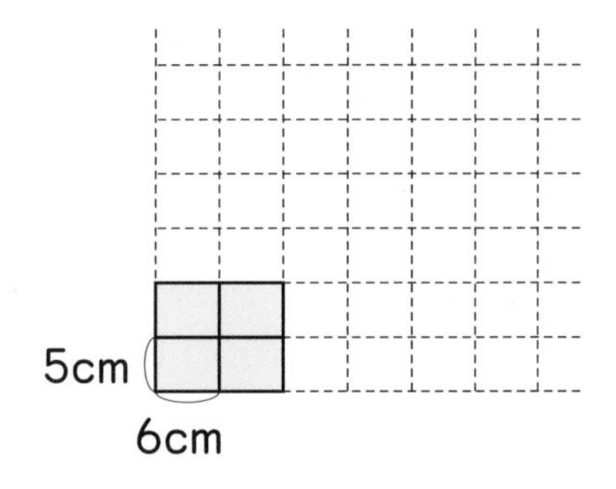

① The length and width became multiple of what number?
② What would be the length in cm of the side of the smallest square created?
③ Let's find out the length of the sides of the first three squares created.

In order to make a square, the length and width should be the same.

3 Boxes of cookies with a height of 6 cm and boxes of chocolates with a height of 8 cm are piled up. What would be the height to make the heights of both piles of boxes the same?

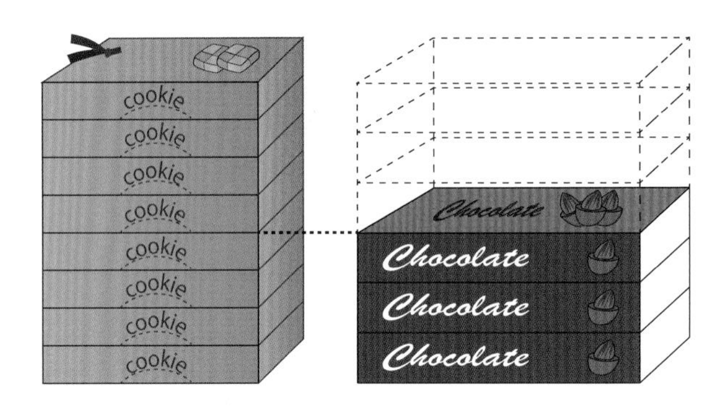

4 There is a metronome that beeps every 12 seconds and another metronome that beeps every 20 seconds. If both start at the same time, how many seconds will it take for them to beep at the same time?

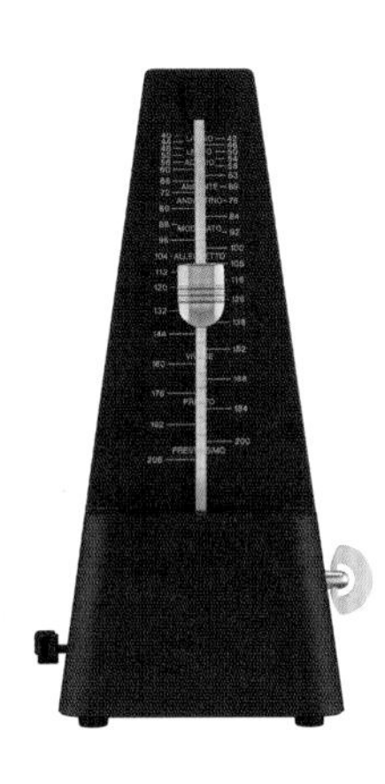

3 Divisors and common divisors

1 **Tessellate squares of the same size so that there are no gaps in a 12 cm long and 18 cm wide rectangle. When you can do this, what is the length of one side of the squares?**

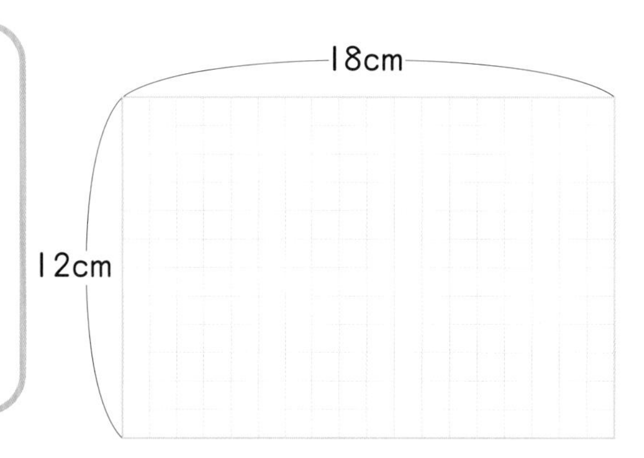

Haruto: A square with a side length of 1cm can be tessellated. But I wonder if we can tessellate other squares...

Akari: First, let's find a square that can be tessellated vertically. And then let's see if we can find a square that can be tessellated horizontally.

Want to think

? Purpose **What is the length of one side of a square when you tessellate squares of the same size inside rectangles without any gaps in between?**

Way to see and think
Let's think separately the horizontal and vertical tessellations.

① What is the length of one side of the squares if you make a vertical tessellation in exactly 12 cm?

In a vertical tessellation of 12 cm without gaps, the possible lengths of one side of the squares are: 1cm, 2cm, 3cm, 4cm, 6cm, and 12cm.

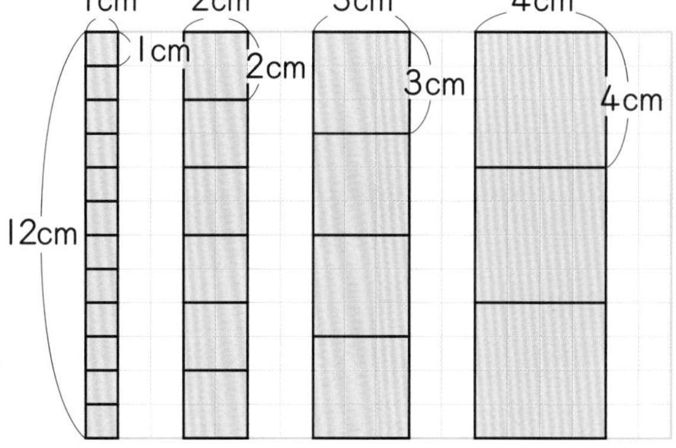

② Let's try to divide 12 by 1, 2, 3, 4, 6, and 12, independently.

If you can divide 12 by 1, 2, 3, 4, 6, and 12 without remainder. Then, those numbers are called **divisors** of 12.

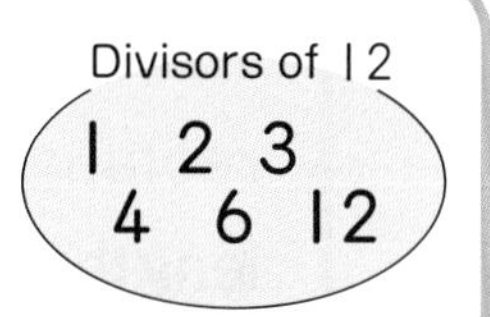

③ What can you tell when the divisors of 12 are paired as shown below?

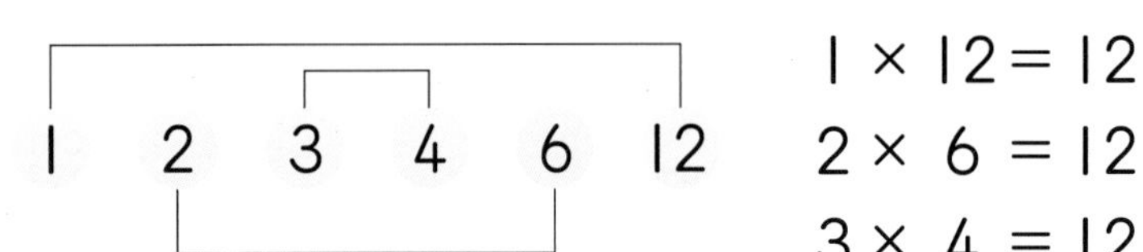

$1 \times 12 = 12$
$2 \times 6 = 12$
$3 \times 4 = 12$

For any whole number, 1 and the number itself are divisors.

④ What is the length of one side of the squares if you make a horizontal tessellation in exactly 18 cm?

In a horizontal tessellation of 18 cm without gaps, the possible lengths of one side of the squares are: 1 cm, 2 cm, 3 cm, 6 cm, 9 cm, and 18 cm.
Then, 1, 2, 3, 6, 9, and 18 are the divisors of 18.

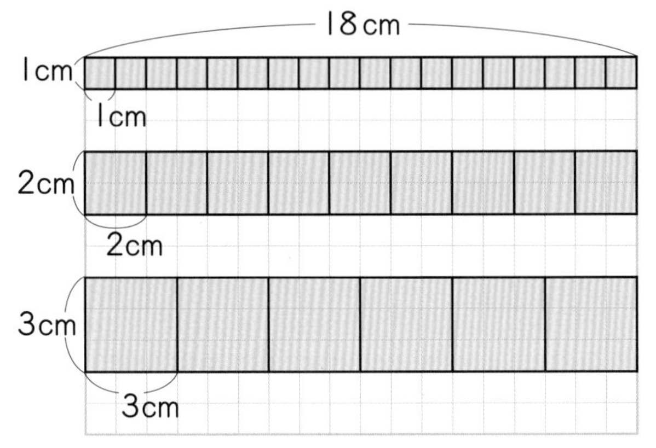

⑤ What is the length in cm of one side of the square if you make a vertical and a horizontal tessellation without gaps?

Vertical ···	1	2	3	4	6	12	(cm)
Horizontal···	1	2	3	6	9	18	(cm)

The numbers that are divisors of both 12 and 18 are called **common divisors**. The greatest of all common divisors is called **the greatest common divisor**.

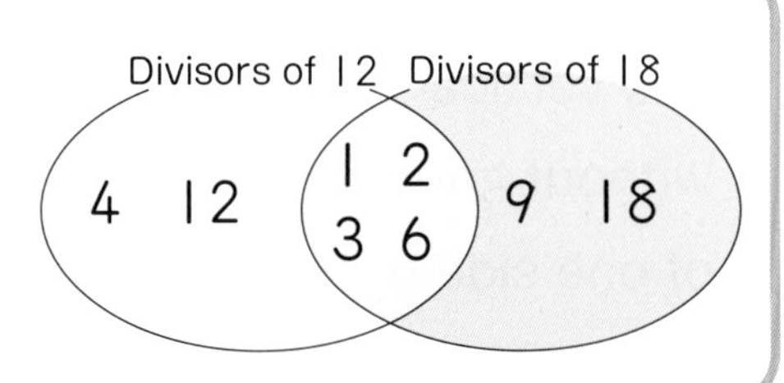

⑥ Let's write all the common divisors of 12 and 18. Also, what is the greatest common divisor between 12 and 18?

The number of the length of one side of a square tessellated into a rectangle without gaps is the common divisor of the number of vertical lengths of the rectangle and the number of horizontal lengths of the rectangle.

1 Let's find all the divisors of 8 and 36. Also, let's find all the common divisors between 8 and 36.

? I wonder if the process to find the common divisor is the same as the common multiple...

2 Let's think about how to find the common divisors of 18 and 24.

Want to think

Purpose How can we find the common divisor?

① Let's explain the ideas of the following children.

Sara's idea

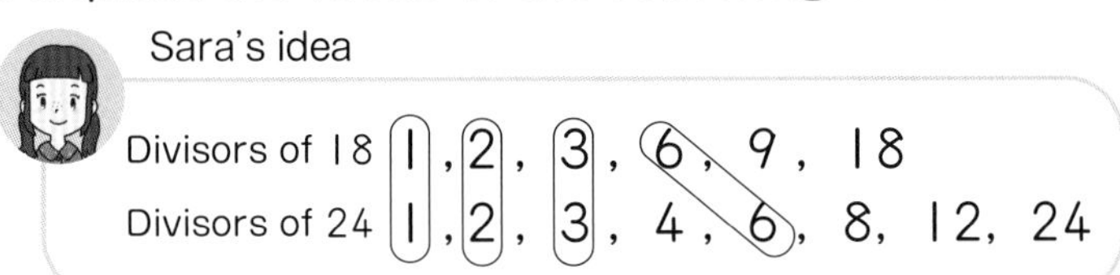

Haruto's idea

Divisors of 18 1, 2, 3, 6, 9, 18

In the case of 18 and 24, you find it from the divisors of 18, the smaller number.

② Let's find all the common divisors of 18 and 24. Then, what is the greatest common divisor?

③ What can you conclude by looking at the common divisors and the greatest common divisor between 18 and 24?

Summary

The common divisors of 18 and 24 are divisors of 6, which is the greatest common divisor of 18 and 24.

1 Let's find all the common divisors of the following pair of numbers. Also, let's find the greatest common divisor between them.

① (8, 16) ② (15, 20) ③ (12, 42) ④ (13, 9)

2 I want to give 8 pens and 12 notebooks, in a way that every child receives the same number of items per type without remainder. To how many children can I give with the given conditions?

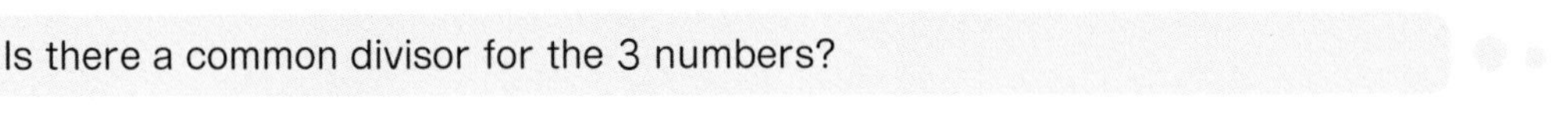

?
Is there a common divisor for the 3 numbers?

3 Let's find the common divisors of 6, 9, and 12.

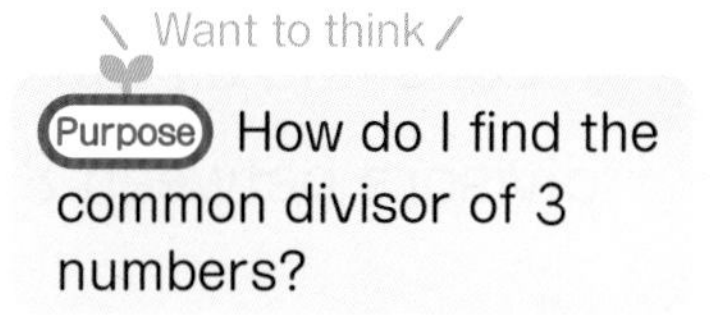

❶ Let's place a ◯ on the divisors of 6, 9, and 12 in each of the lines on the right.

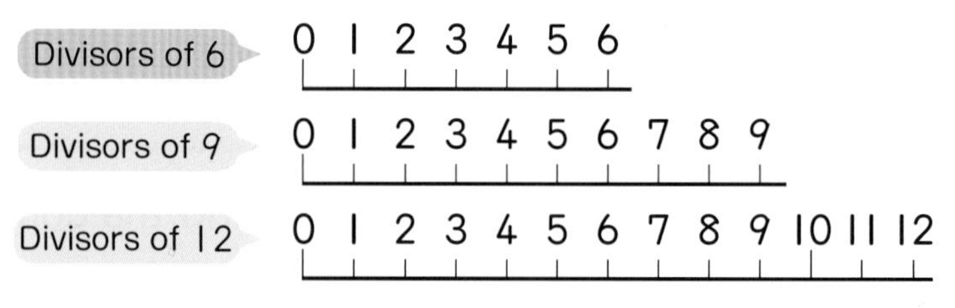

❷ Let's find all the common divisors of 6, 9, and 12. After, find the greatest common divisor.

1 Let's find all the common divisors of the following set of numbers. Also, let's find the greatest common divisor.

① (8, 18, 24)　② (6, 12, 27)　③ (7, 16, 23)

Summary The common divisor of 3 numbers can be found in the same way as the common divisor of 2 numbers.

2 Let's explore about the relationship between multiples and divisors.

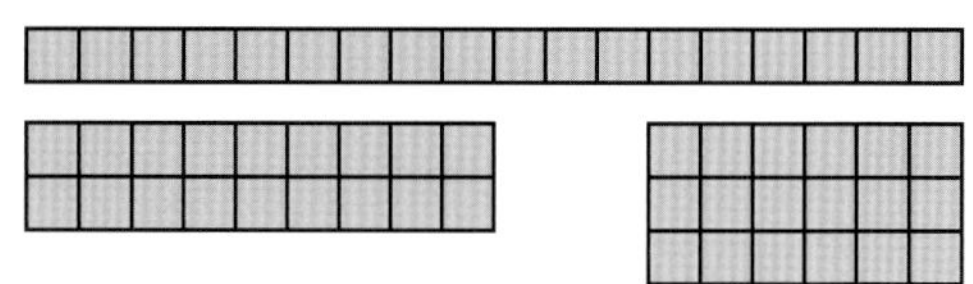

① Let's align 18 square cards in a rectangular shape and find the divisors of 18.

② Is 18 a multiple of the divisors found in ①?

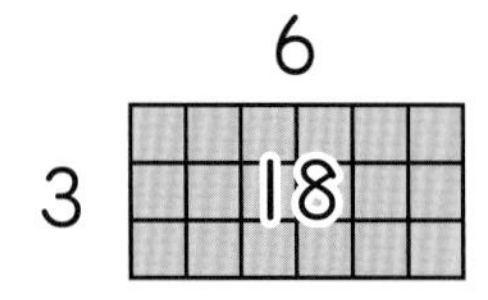

- 3 and 6 are divisors of 18.
- 18 is a multiple of 3 and 6.

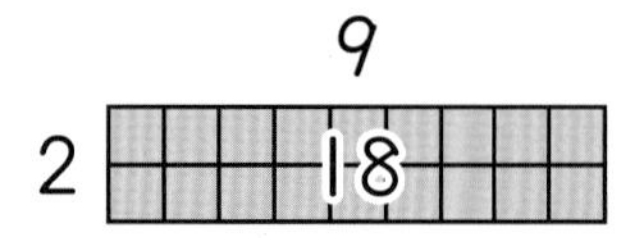

- 2 and ☐ are divisors of 18.
- 18 is a multiple of 9 and ☐.

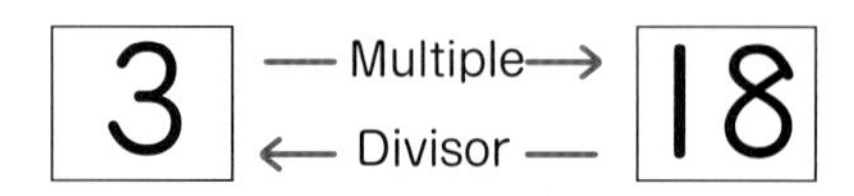

I can say that 18 is a multiple of 3, and 3 is a divisor of 18.
Akari

CAN What can you do?

☐ We understand about even and odd numbers. → pp.57 ~ 58

1 Let's classify the following whole numbers as even or odd numbers.

0 1 2 4 8 9 13 17 68 108 493 78542

Even number	Odd number

☐ We can find multiples and divisors. → pp.60 ~ 61, pp.67 ~ 68.

2 Let's find the first 3 multiples of the following numbers. Also, let's find all the divisors.

① 6 ② 13 ③ 16 ④ 24

☐ We can find common multiples and the least common multiple. → pp.62 ~ 63

3 Let's find the first 3 common multiples of the following set of numbers. Also, let's find the least common multiple.

① (3, 6) ② (5, 7) ③ (6, 10) ④ (8, 12)

☐ We can find common divisors and the greatest common divisor. → pp.68 ~ 69

4 Let's find all the common divisors of the following set of numbers. Also, let's find the greatest common divisor.

① (9, 15) ② (12, 24) ③ (30, 42) ④ (28, 42)

☐ We can solve problems using common multiples and common divisors. → p.66

5 At a station, a train departs every 12 minutes and a bus departs every 8 minutes. At 9 am, the train and the bus departed at the same time. When would be the next time when they both depart together?

Supplementary Problems → p.159

Which "Way to See and Think Monsters" did you find in "5 Multiples and Divisors"?

I found "Summarize" when I was trying to classify the whole numbers.

I found other monsters, too!

Utilize Usefulness and Efficiency of Learning

1 I would like to use a balance to weigh different objets. However, when weighing, I can only use two types of weights: 3 g weight and 10 g weight. Also, the weight can be placed only on the right dish. What kind of weight can you weigh?

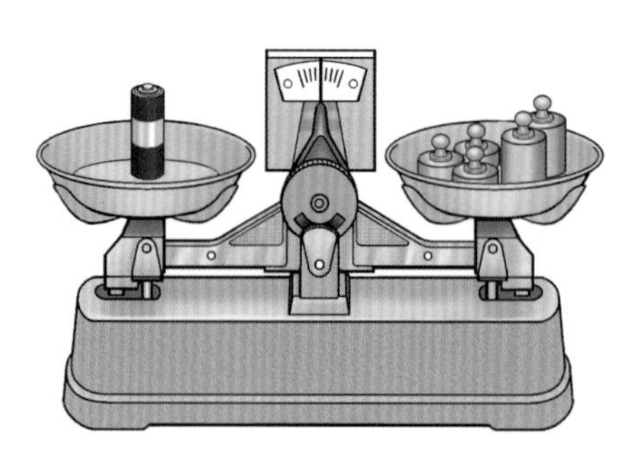

Haruto

I can weigh multiples of 3 and multiples of 10.

Any other weight different than that, I cannot weigh it.

Sara

① When a battery was weighed, three 3 g weights and two 10 g weights were used to balance. Therefore, what is the weight of the battery?

② Which of the following weights Ⓐ~Ⓓ cannot be weighed with this balance?

Ⓐ 9 g　　Ⓑ 10 g　　Ⓒ 11 g　　Ⓓ 12 g

③ Is there any other weight that cannot be weighed? Let's investigate using the table on the right by placing a ○ on the weights you can weigh and by placing a × on the weights you cannot weigh.

1	2	3	4	5	6	7	8	9	10
11	12	13	14	15	16	17	18	19	20
21	22	23	24	25	26	27	28	29	30
31	32	33	34	35	36	37	38	39	40
41	42	43	44	45	46	47	48	49	50
51	52	53	54	55	56	57	58	59	60
61	62	63	64	65	66	67	68	69	70
71	72	73	74	75	76	77	78	79	80
81	82	83	84	85	86	87	88	89	90
91	92	93	94	95	96	97	98	99	100

Yu

It is not easy to check one by one.

④ Using the idea of multiples, let's try to organize which numbers can be weighed.

With the Way to See and Think Monsters...

Let's Reflect!

Let's reflect on which monster you used while learning "5 Multiples and Divisors."

Summarize

Whole numbers can be summarized in terms of even and odd numbers, or in terms of multiples and divisors.

① What kind of numbers are even numbers and odd numbers?

Akari: An even number is a number divisible by ☐. It can also be seen as a number multiplied by 2.

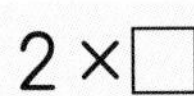

$2 \times \square$

Even number

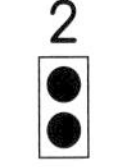

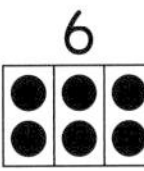
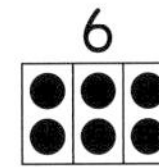

2 4 6

Haruto: An odd number is a number that is not divisible by 2. It can also be seen as a number multiplied by 2 and add ☐.

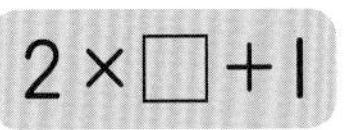

$2 \times \square + 1$

Odd number

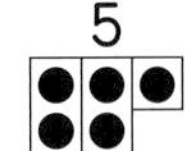
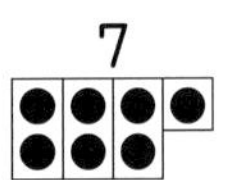

3 5 7

② Let's summarize multiples and divisors.

Sara: The number obtained from the multiplication between any number and a whole number, such as 1, 2, 3, ..., is called a multiple of the number. Also, when you can divide a whole number by another whole number without remainder, that number is called its divisor.

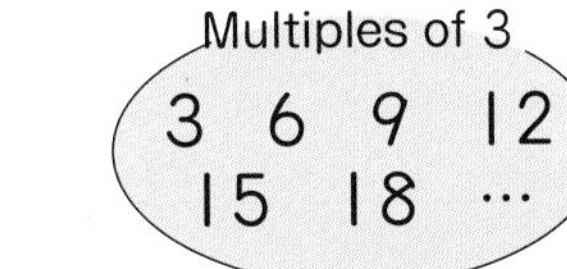

Divisors of 12

1 2 3
4 6 12

Yu: 3 x 4 = 12, so 12 is a multiple of 3 and 3 is a divisor of 12.

? Solve the ?

Whole numbers can be divided into either even or odd numbers. They can also be viewed as multiples or divisors.

Yu

Want to Connect

Can we summarize whole numbers in other ways?

Haruto

Reflect Connect

Problem

Number 8 ______________ 2 ______________ .

Before and after the digit 2, there can be words, numbers or blank spaces.

I found the message shown above incomplete. I wonder what was written on it.

Which words and numbers do you think can complete the sentence?

Purpose Let's try to think different ways to represent numbers.

Number 8 is 2 added four times. (Addition)

(Math Equation) $8 = 2 + 2 + 2 + 2$

↔ Same meaning

Number 8 is 2 grouped four times.

(Math Equation) $8 = 2 \times 4$ (Multiplication)

Number 8 is 2 subtracted from 10.

(Math Equation) $8 = 10 - 2$ (Subtraction)

Area of a rectangle

↓

Number 8 is 2 multiplied by 4.

Number 8 is 4 times of 2 ___ .

Number 8 is 2 multiplied by itself three times

(Math Equation) $8 = 2 \times 2 \times 2$ (Multiplication)

Can you find ways to represent number 8 using the number 2?

Haruto

If you change the math equation $8 = 2 \times 4$ into words, it becomes "number 8 is 4 times 2."

Sara

If 2×4 is represented by a figure, it looks like this?

Akari

When represented in an equation, it won't work for division.

Is there any way to represent it using division?

(Math Equation) $8 = 16 \div 2$ (Division)

↓

Number 8 is 16 divided by 2____.

Number 8 is 2 larger than 6.

(Math Equation) $8 = 2 + 6$ (Addition)

Other ways:

Number 8 is a multiple of 2____.

Multiples of 2:

2, 4, 6, 8……

Even number ⟷ Odd number

0 is also an even number.

Using "common multiple":

Number 8 is a common multiple of 2 and 4.

Using "divisor":

Number 8 is a divisor of 24.

Summary You can think different ways to represent numbers.

- One number can be represented by the sum, difference, product, or quotient of other numbers.
- Multiplication is performed when a number is represented as the area of a rectangle or square.

  ↓

  1, 4, 9, 16...... (Product of the same 2 numbers)

- If you use multiples of divisors, it will come as member of the same category inside whole numbers.

The area of a 2 cm long and 4 cm wide rectangle, $2 \times 4 = 8$, is $8\ cm^2$.

The number of cubic blocks is....

Want to Connect

If 17 is represented as a product, the only possible way is 1×17. Are there any other numbers like this?

It will continue in Junior High School.

Let's deepen. → p.168

Find the ? Which is more crowded?

\ Want to know /

Purpose What should we do to compare the crowdedness of people?

Measure per Unit Quantity (1)

Let's think how we can change the unit for comparing.

Crowdedness →

1 Children are sitting on picnic mats. Let's explore which of the following Ⓐ, Ⓑ or Ⓒ is more crowded?

Number of children on mats

	Number of mats	Number of children
Ⓐ	4	12
Ⓑ	6	15
Ⓒ	6	12

Ⓐ 12 children sitting in 4 mats.

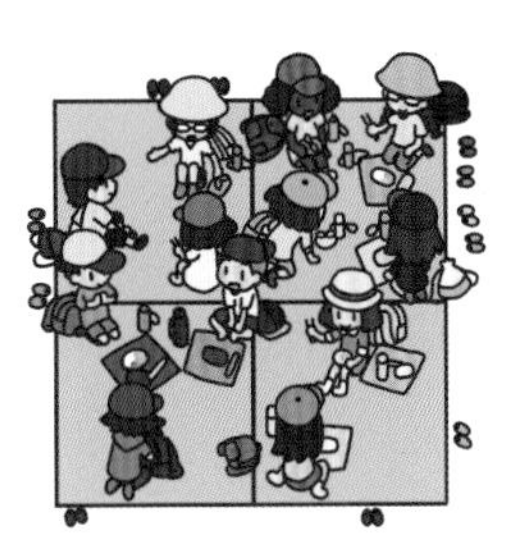

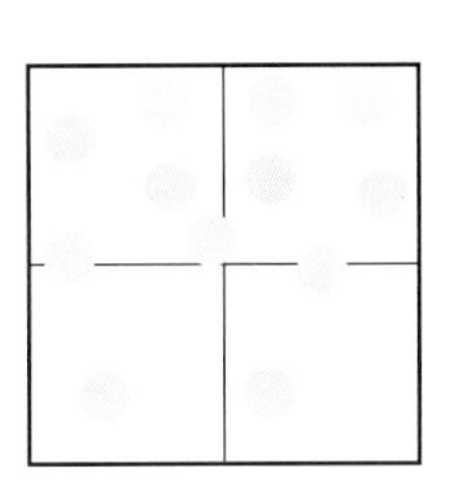

Ⓑ 15 children sitting in 6 mats.

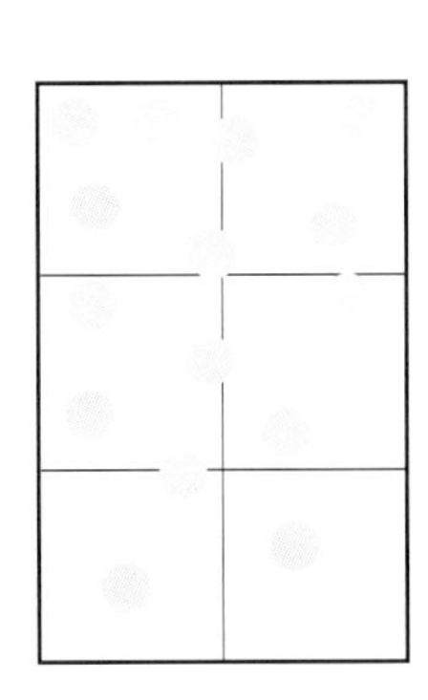

Ⓒ 12 children sitting in 6 mats.

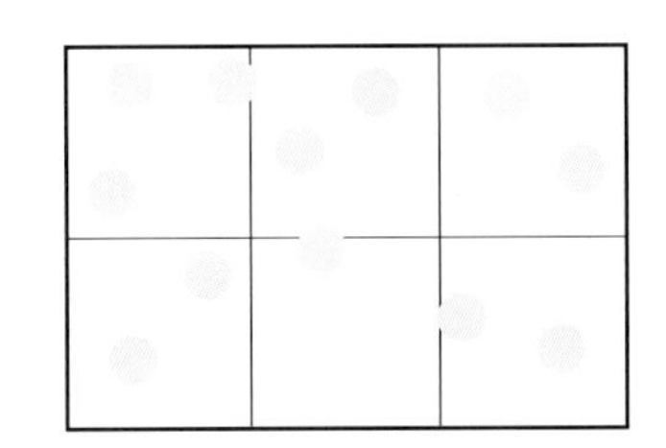

There are more people in Ⓑ.

I wonder if it is more crowded because Ⓐ has the least number of mats...

❶ Between Ⓐ and Ⓒ, which one is more crowded?

Ⓐ 12 children in 4 mats　　Ⓒ 12 children in 6 mats

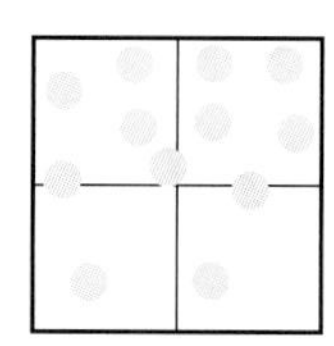

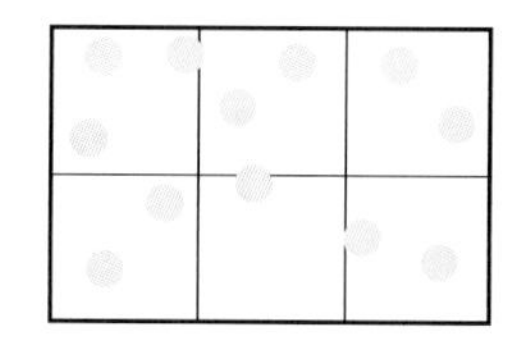

Way to see and think

A comparison can be made if you have either the same number of people or the same number of mats.

❷ Between Ⓑ and Ⓒ, which one is more crowded?

Ⓑ 15 children in 6 mats　　Ⓒ 12 children in 6 mats

❸ Now, between Ⓐ and Ⓑ, which one is more crowded? Let's explain and compare the ideas of the following children.

Way to see and think

Compare the number of children on one mat using the mean idea.

Sara's idea

Comparing how many children are on one mat.

Ⓐ $12 \div 4 = 3$

3 children per mat.

Ⓑ $15 \div 6 = 2.5$

2.5 children per mat.

Since there are more people in 1 mat, we can say that ☐ is more crowded.

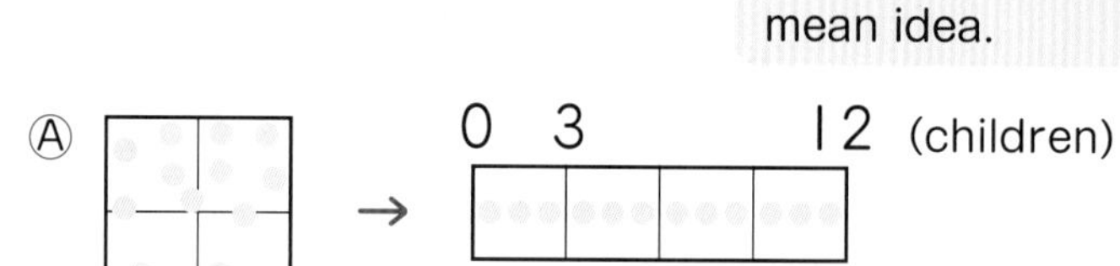

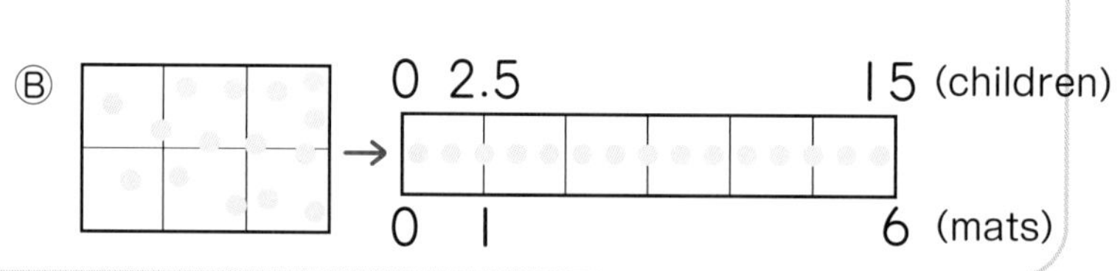

Way to see and think

Compare the area occupied per child.

Haruto's idea

Comparing how much of a mat is occupied by one child.

Ⓐ $4 \div 12 = 0.333\cdots$　　About 0.33 of the mat per child.

Ⓑ $6 \div 15 = 0.4$　　0.4 mat per child.

Since the area occupied per children is less, we can say that ☐ is more crowded.

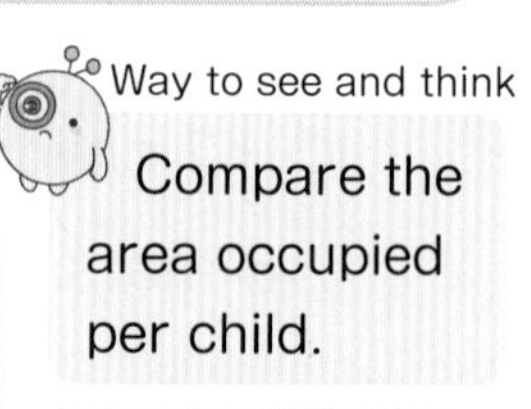

That means the area is more narrow for one person.
Akari

Yu's idea

Comparing the number of children on 12 mats.

Ⓐ $12 \times 3 = 36$ 36 children on 12 mats.

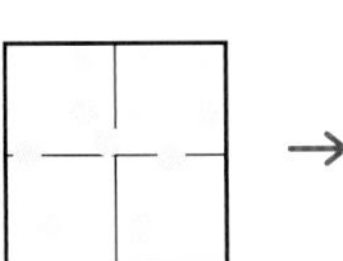
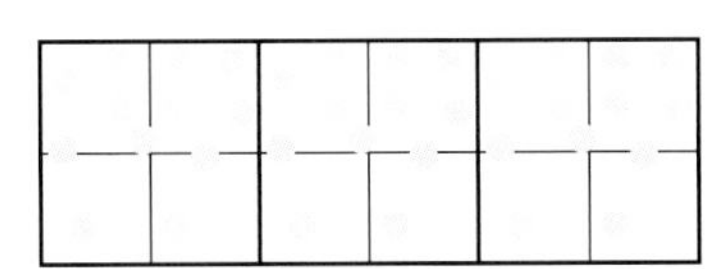

Ⓑ $15 \times 2 = 30$ 30 children on 12 mats.

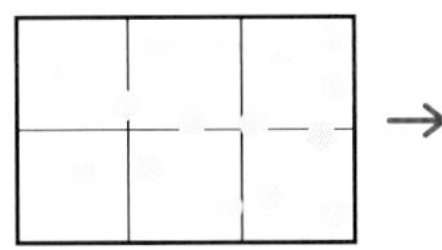
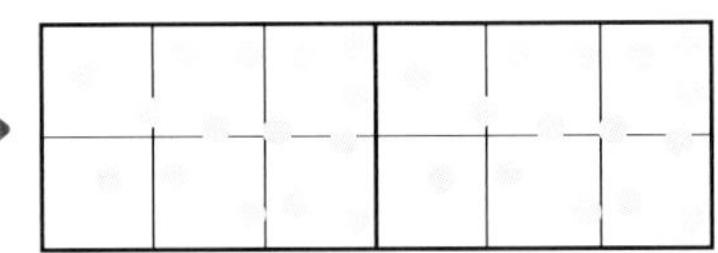

When 12 mats are arranged, we can say that is more crowded [ ].

Way to see and think

Yu are using the common multiples of 4 and 6 to arrange and compare the number of mats.

Summary

We can compare crowdedness by considering the number of people per mat, the area per person, or by making the number of mats equal.

? What kind of comparisons should we make so that we can compare at any time?

2 There are 32 mats in the gymnasium and 96 children are about to sit on them. When all of them are seated, let's compare the results with Ⓐ, Ⓑ and Ⓒ from 1.

Number of children on mats

	Number of mats	Number of children
Ⓓ	32	96

\ Want to think /

Purpose What method is easy to understand when comparing crowdedness?

Which of the methods considered in 1 is easier to understand?

Sara

If we use the idea of common multiple, we need to think about each situation.

Haruto

How about finding out the area per person ...

Yu

① If the area of one mat is $1\,m^2$. Then, on Ⓐ, Ⓑ, Ⓒ, and Ⓓ, how many children are per $1\,m^2$?

Ⓐ $12 \div 4 = \square$

Ⓑ $15 \div 6 = \square$

Ⓒ $12 \div 6 = \square$

Ⓓ $96 \div 32 = \square$

Number of children / Area (m^2) / Number of children per $1\,m^2$

Summary

It is easier to understand how to compare crowdedness by finding the number of people per square meter.

The bigger the number, the more crowded it is.
Sara

Usually, crowdedness of people is compared by matching the same unit, such as $1\,m^2$ or $1\,km^2$.

1 There are 12 dogs in a plaza of $80\,m^2$. Next to it, there are 15 dogs in a plaza of $120\,m^2$. Which plaza is more crowded?

2 There is a train with 8 cars carrying 1440 passengers. While another carries 1890 passengers in 10 cars. Which train is more crowded?

3 **The table on the right represents the population and area of East City and West Town. Let's find the number of people per $1\,km^2$ and compare how crowded is each place.**

Population and area

	Population	Area (km^2)
East City	273600	72
West Town	22100	17

The number of people per $1\,km^2$ is called **population density**. The crowdedness of people living in a country or a prefecture is represented by the population density.

1 Choose 3 prefectures from below and calculate the population density of each of them. Let's round to the nearest tenths and give the answer in whole numbers.

Want to try

Purpose Let's find the population density of different prefectures.

Population in 2020

Let's explore the answer for your own town.

Way to see and think
Population density is considered using the same area unit.

Hokkaido
$83424 km^2$
5,228,885 people

Aomori
$9646 km^2$
1,238,730 people

Niigata
$12584 km^2$
2,202,358 people

Hiroshima
$8480 km^2$
2,801,388 people

Osaka
$1905 km^2$
8,842,523 people

Tokyo
$2194 km^2$
14,064,696 people

Fukuoka
$4987 km^2$
5,138,891 people

Shizuoka
$7777 km^2$
3,635,220 people

Kumamoto
$7409 km^2$
1,739,211 people

Kagawa
$1877 km^2$
951,049 people

Kochi
$7104 km^2$
692,065 people

Okinawa
$2283 km^2$
1,468,410 people

Kagoshima
$9187 km^2$
1,589,206 people

? What else can we compare using the same idea?

4

A wire is 8 m long and weighs 480 g. Find the weight of the wire per meter.

I would like to use a diagram to think better...

I would prefer to use a table....

❶ Using diagrams and tables, the math equations to find the weight per meter was considered. Let's explain and compare the ideas of the following children.

Sara's idea

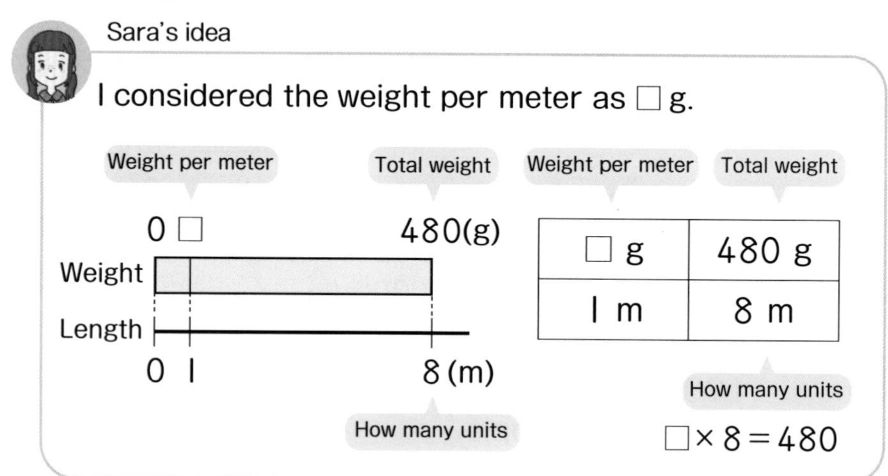

Way to see and think
Using the idea of "measurement per unit" × "how many units" = "total measurement."

Yu's idea

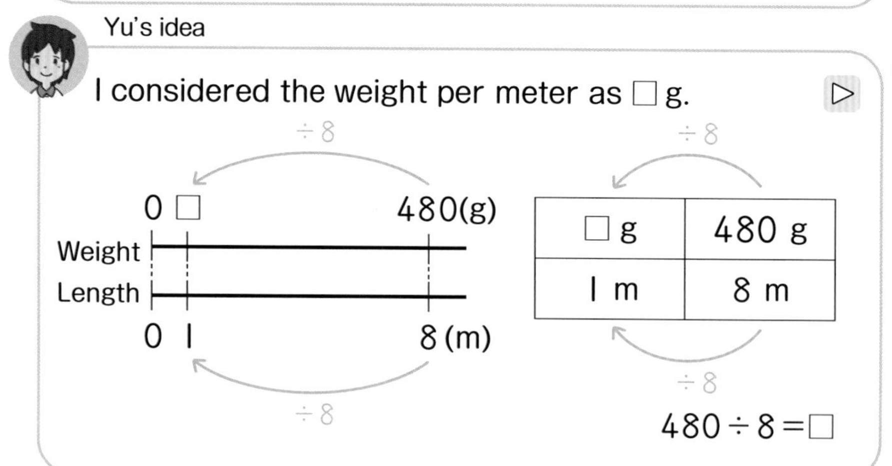

Way to see and think
Using the proportion idea.

Both are using 2 number lines.

❷ What is the weight, in grams, per meter?

\ Want to think /

If we know the weight per meter, what else can we find?

③ What is the weight, in grams, of a 15 m long wire? Let's think by using a diagram and a table. ▷

15 m is 15 times 1 m, so the weight is also 15 times 60 g.
Akari

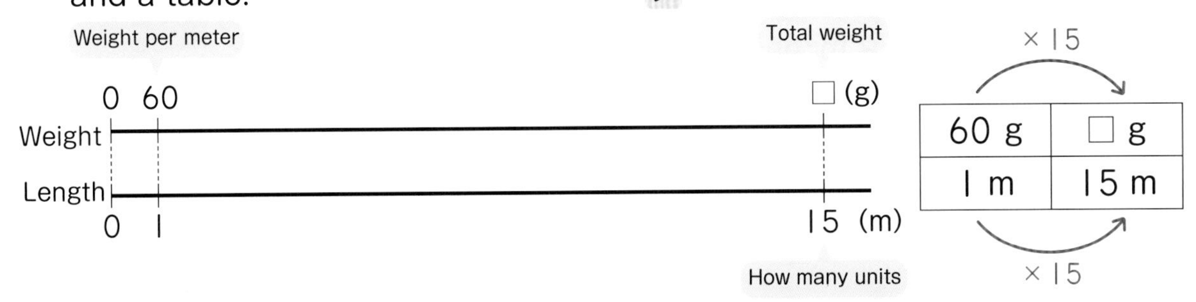

④ This wire was cut and weighed 300 g. What is the length, in meters, of this piece of wire? ▷

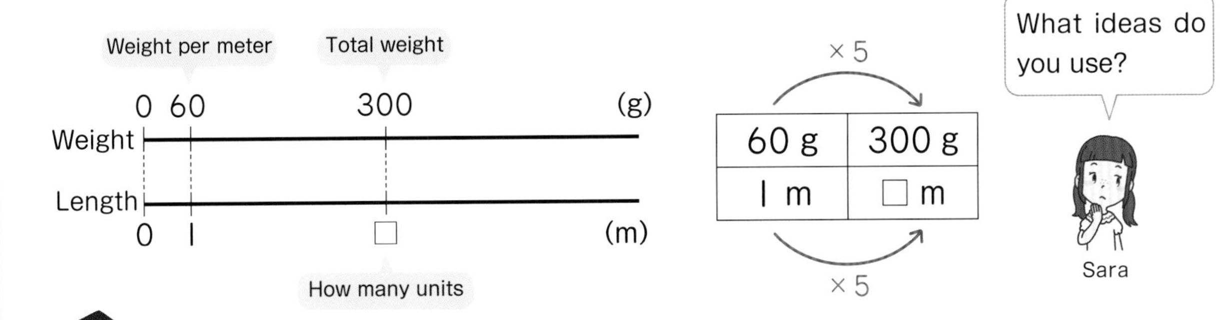

Population density, weight per meter, etc. are called **measure per unit quantities**.

Summary

Once you know the measure per unit quantity, you can use diagrams, tables, and ideas such as proportion to find how many units and the total measurement.

The measure per unit quantity is what you've been thinking of as "number for each."

What does this have to do with the multiplication and division we have learned until now?

Let's think about it on page 90.

1 A wire is 15 m long and weighs 600 g. What is the weight, in grams, of the wire per meter?

? In what other situations can the measure per unit quantity be used?

5

We harvested potatoes at school. Field A was 6 m^2 and the total weight of potatoes harvested was 43.2 kg. Field B was 9 m^2 and the total weight of potatoes harvested was 62.1 kg. Which field produced more potatoes? Let's compare by using the weight per 1 m^2.

\ Want to try /

Purpose Let's consider various problems using measure per unit quantity.

Akari

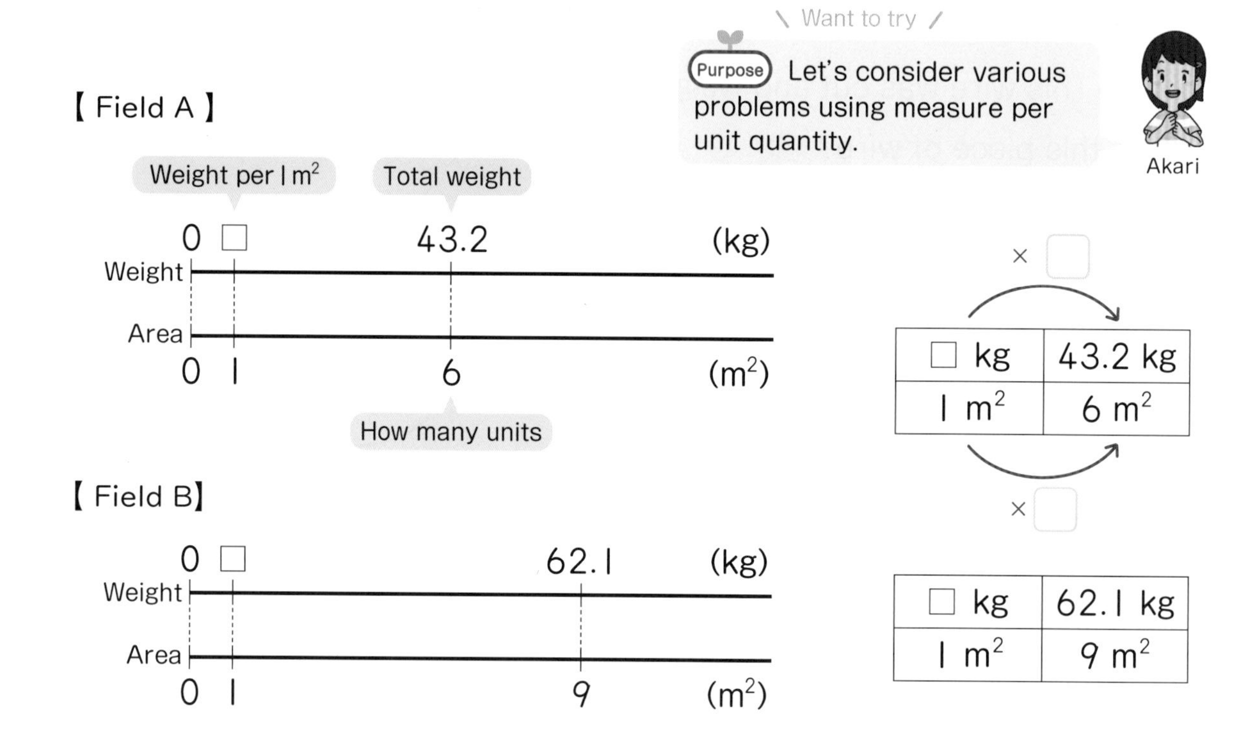

□ kg	43.2 kg
1 m^2	6 m^2

□ kg	62.1 kg
1 m^2	9 m^2

1 There are 10 notebooks that cost 1200 yen in total and 8 notebooks that cost 1040 yen in total. Which notebook is more expensive? Let's think about the cost per notebook.

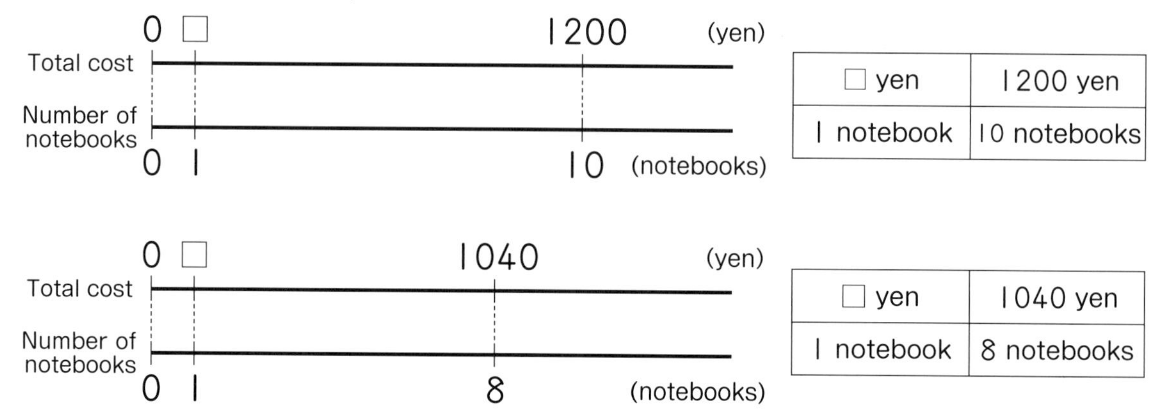

□ yen	1200 yen
1 notebook	10 notebooks

□ yen	1040 yen
1 notebook	8 notebooks

2 Copy machine A can make 300 copies per 4 minutes and copy machine B can make 532 copies per 7 minutes. Which copy machine prints more copies per minute?

【A】

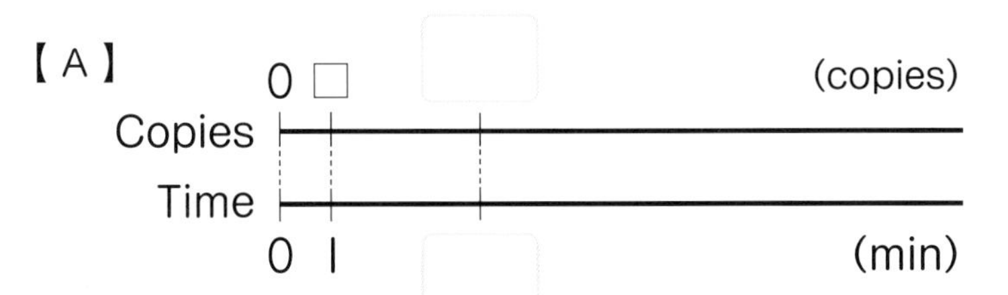

□ copies	copies
1 min	min

【B】

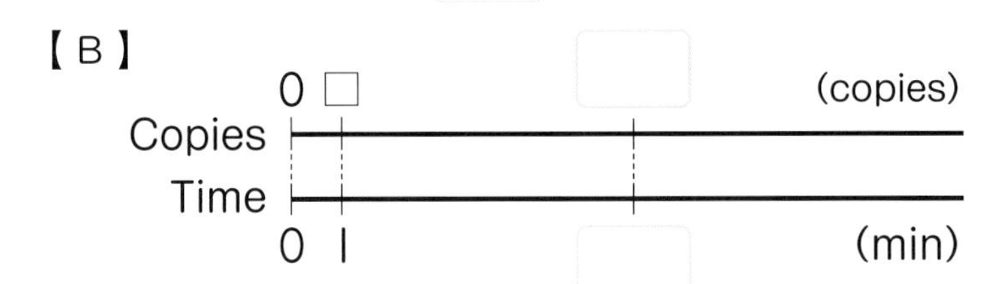

□ copies	copies
1 min	min

3 There is a car that runs 720 km with 45 L of gasoline. Let's answer the following questions about this car.

① Let's find the distance that runs per liter of gasoline.

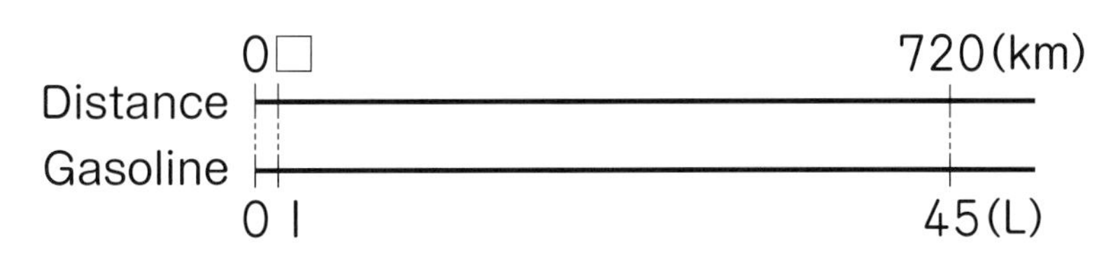

□ km	
1 L	

② What distance, in km, will the car run with 32 L of gasoline?

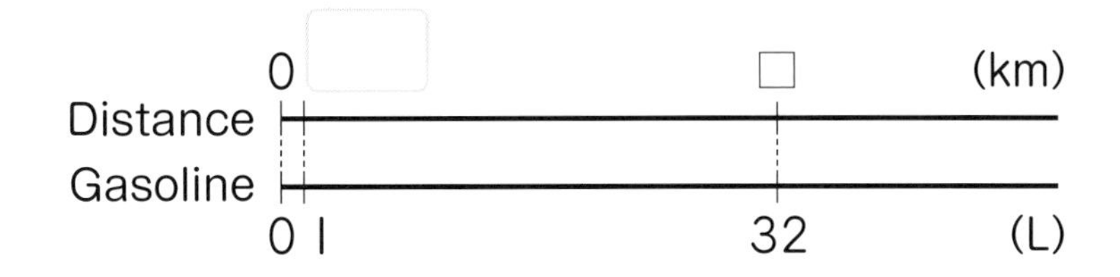

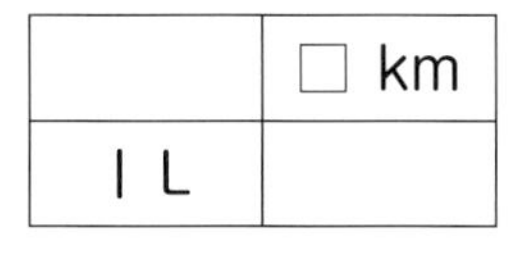

	□ km
1 L	

③ To run 1024 km, how many liters of gasoline are needed?

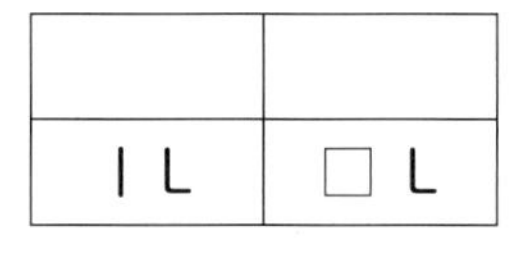

1 L	□ L

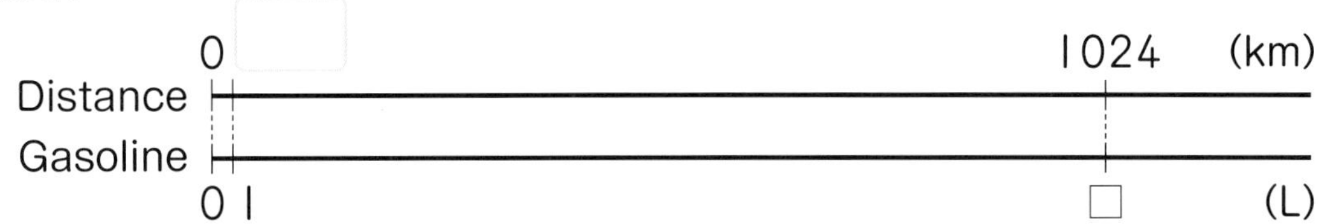

CAN What can you do?

☐ We can compare crowdedness. → pp.79 ~ 80

1 Between Ⓐ and Ⓑ, which one is more crowded?

① Ⓐ 17 people in 6 mats. Ⓑ 15 people in 4 mats.

② Ⓐ a train with 840 passengers in 6 cars

Ⓑ a train with 1000 passengers in 8 cars.

☐ We can find the population density. → pp.80 ~ 81

2 The population and area of two cities and one town are shown in the table on the right. Answer the following questions.

① Find the population density of North City and South City in whole numbers, rounding to nearest tenth.

② Arrange the population density in descending order.

Population and area

	Population	Area (km^2)
North City	267200	70
South City	185000	47
East Town	58000	14

☐ We can find and compare measure per unit quantities. → pp.82 ~ 85

3 Let's answer the following questions.

① 12 colored pencils cost 600 yen. While, 8 colored pencils cost 440 yen. Which colored pencil is more expensive? Let's compare using the price per pencil.

② A wire is 5 m long and weighs 370 g. Let's consider this wire.

Ⓐ Let's find the weight per meter.

Ⓑ What would be the weight, in grams, for a 10 m long wire?

Ⓒ A wire of the same type was measured and weighed 555 g. What would be the length, in meters, of this wire?

③ In factory A, 8 workers put 120 products in a box in one hour. While, in factory B, 12 workers put 192 products in a box in one hour. Which factory packs more boxes per worker?

Supplementary Problems → p.160

Which "Way to See and Think Monsters" did you find in " 6 Measure per Unit Quantity (1)" ?

I found "Align" when I was comparing the crowdedness.

Haruto

When I was thinking about the measure per unit quantity...

Sara

Utilize Usefulness and Efficiency of Learning

1 As I investigated global warming, I heard that one of the causes is the increased carbon dioxide in the air. So, I checked how much carbon dioxide is emitted in Japan and found the following table. Answer the following questions.

Amount of carbon dioxide emitted

Year	Emissions of carbon dioxide (ten thousand kg)	Population (ten thousand people)	Emissions of carbon dioxide per person (kg)
1994	123200000	12527	
1999	124600000	12667	
2004	128600000	12779	
2009	116500000	12803	
2014	126500000	12724	
2019	110800000	12617	

① How does it change per person? Let's try to think and represent it on a bar graph or a line graph.

② The following graph shows the amount of carbon dioxide emissions per person in different countries. What can you say?

Carbon dioxide emissions per person in different countries (2018)

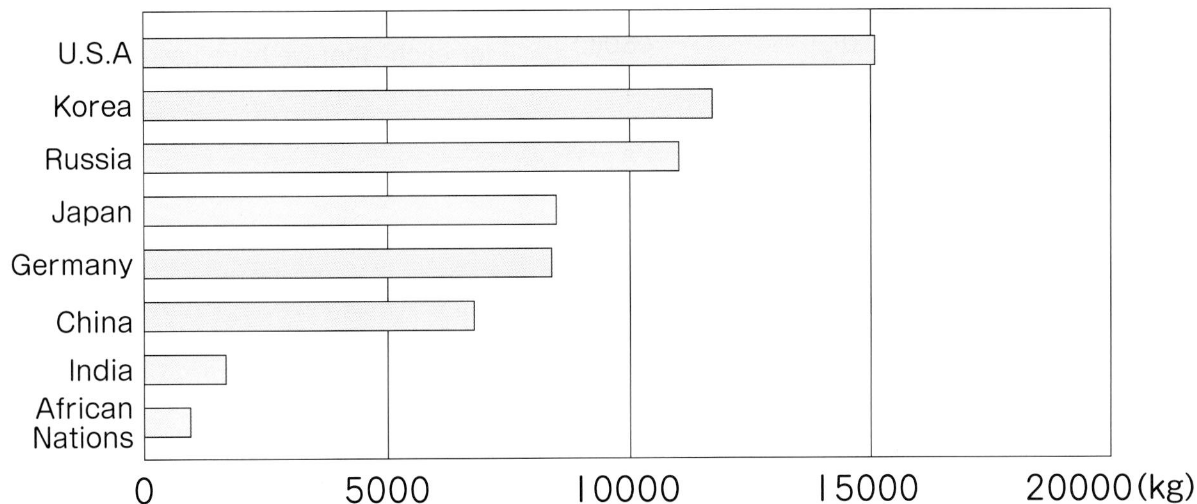

With the Way to See and Think Monsters...

Let's Reflect!

Let's reflect on which monster you used while learning " 6 Measure per Unit Quantity (1)."

Align

By making same the number of people and the number of mats, we are able to compare the crowdedness.

① Which one is the most crowded Ⓐ, Ⓑ, or Ⓒ? Explain your ideas to find out the answer.

Number of children on mats

	Number of mats	Number of children
Ⓐ	4	12
Ⓑ	6	15
Ⓒ	6	12

I thought about how many people are on one mat.
Ⓐ $12 \div 4 = 3$
Ⓑ $15 \div 6 = 2.5$
Ⓒ $12 \div 6 = 2$

Sara

I thought about how much space of the mat one children occupies.
Ⓐ $4 \div 12 = 0.33...$
Ⓑ $6 \div 15 = 0.4$
Ⓒ $6 \div 12 = 0.5$

Haruto

Since the least common multiple between 4 and 6 mats is 12, then I thought the proper number of mats should be 12.

Yu

Unit

Measure per unit quantity is the same as thinking the number for each.

② What was the number that represented the measurement per unit quantity?

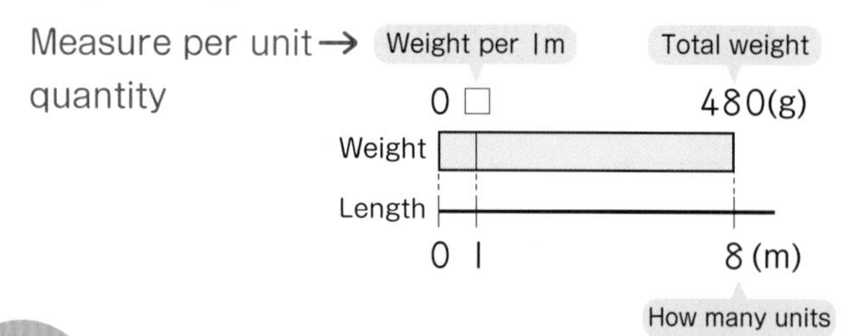

The idea of measure per unit quantity is the same as "number for each" that we have used in multiplication and division so far.

Akari

Solve the ?

It was easy to compare crowdedness by using the measure per unit quantity.

Akari

Want to Connect

Are there other things we can think by using the measure per unit quantity?

Yu

Math Patrol

**When comparing crowding, we sometimes compare by finding the size equivalent to one portion.
In this case, it is necessary to pay attention to the size of one "of what" is being compared.**

① The table on the right shows the area of mats Ⓐ and Ⓑ and the number of people on them. To find out which mat is more crowded, we did the following calculations.

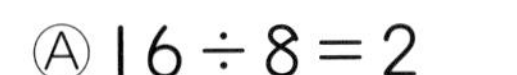

Ⓐ $16 \div 8 = 2$ Ⓑ $9 \div 5 = 1.8$

What did you get from the calculation above? Choose one of the following ⓐ~ⓓ and answer its symbol.

Area of the mat and number of people on it

	Area(m^2)	People
Ⓐ	8	16
Ⓑ	5	9

ⓐ Since the number of people per $1\ m^2$ is 2 and 1.8, Ⓐ is more crowded.

ⓑ Since the number of people per $1\ m^2$ is 2 and 1.8, Ⓑ is more crowded.

ⓒ Since the area per person is $2\ m^2$ and $1.8\ m^2$, Ⓐ is more crowded.

ⓓ Since the area per person is $2\ m^2$ and $1.8\ m^2$, Ⓑ is more crowded.

Since both equations are divided by the area, we are looking for the ....

Sara

Frequently found mistake
Think of it as an equation for the area per person and choose ⓒ or ⓓ.

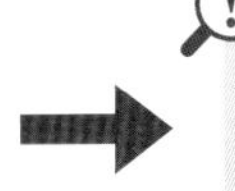

Be careful!
Let's decide if we divide by the area or the number of persons. In the above equations Ⓐ and Ⓑ, the divisor is the area and you will find the number of persons per m^2 so the answer is either ⓐ or ⓑ. If the number of people per $1\ m^2$ is higher, it means it is more crowded, so the answer is Ⓐ.

② A room has an area of $8\ m^2$ and there are 16 persons in the room. Write a math equation to find the number of persons per m^2 in this room.

The number of people per $1\ m^2$ is required, so the divisor should be...

Haruto

Frequently found mistake
Do $8 \div 16$.

Be careful!
To find the number of persons per m^2, divide the number of persons by the area.

Be careful the questions asks for the size of one "of what." The way to find it depends on whether you want to find the number of persons per $1\ m^2$ or the area per person.

Reflect

Connect

Problem Let's try to solve various problems.

① A string of 12 m is divided into 3 equal parts. What is the length of one part in meter?

Math Equation: $12 \div 3 = 4$

Answer 4 m

Representation in a table

□ m	12 m
1 string	3 strings

The same as measure per unit quantity.

1 string unit = per string

② The area for one sheet of origami is 225 cm^2. What is the area covered, in cm^2, by 4 sheets of origami?

Math Equation: $225 \times 4 = 900$

Answer 900 cm^2

225 cm^2	□ cm^2
1 sheet	4 sheets

Similar to the problem on page 83

All of these were learned in the 3rd grade. However, it is similar to the measure per unit quantity that we are learning now.

Akari

Haruto

1 string unit = per string

1 sheet unit = per sheet

right?

③ There is 1 L 5dL of juice. If every time you drink 3 dL, how many times can you drink?

1 L 5dL = 15 dL

Math Equation: $15 \div 3 = 5$

Answer 5 times

3 dL	15 dL
1 time	□ times

The same as measure per unit quantity

If the units are aligned with L...

1 L 5dL = 1.5 L

3 dL = 0.3 L

$1.5 \div 0.3 = ?$

Summary

Until now, multiplication and division have used the idea of measure per unit quantity.

Same Way

Yu

The multiplications and divisions learned so far were related to the measure per unit quantity.

Want to Connect

Sometimes the divisor is a decimal number. I wonder if we can still calculate...

Sara

Let's think about the problem by using diagrams and tables.

When you don't know the total number

A wire has a weight of 5.8 g per meter. What is the weight of 5 m of this wire?

If you draw a diagram and think in terms of word formulas....

Akari

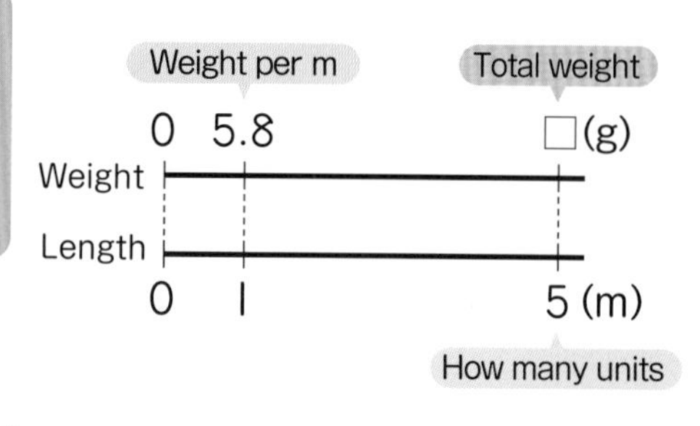

Weight per m × How many units = Total weight

Therefore, $5.8 \times 5 = 29$

Answer: 29 g

Let's draw a table and think about it using the idea of proportionality...

Haruto

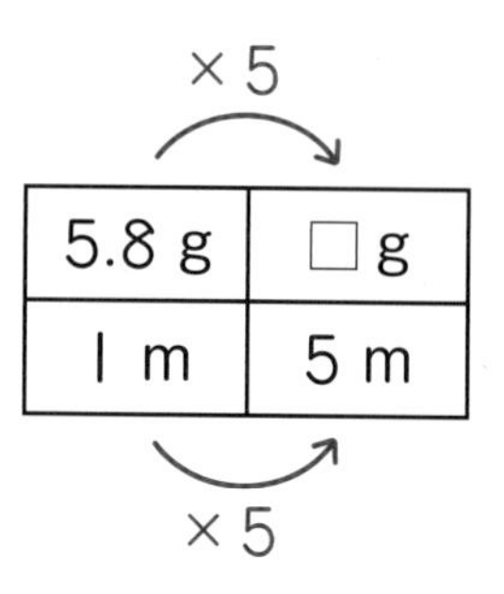

5.8 g	□ g
1 m	5 m

$5.8 \times 5 = 29$

Answer: 29 g

How to draw a diagram ▷

(1) Draw straight lines for weight and length.

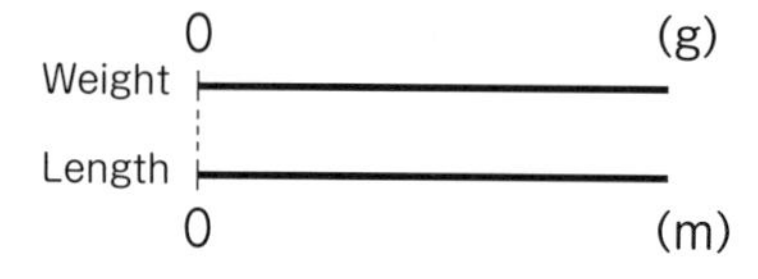

(2) Draw a mark for 1 m and for 5.8 g in the corresponding place.

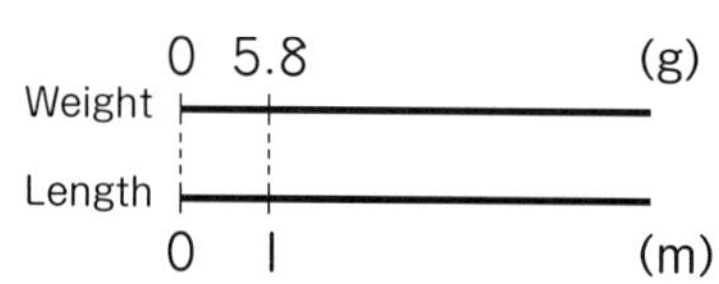

(3) Draw a mark for 5 m and □ g in the corresponding place.

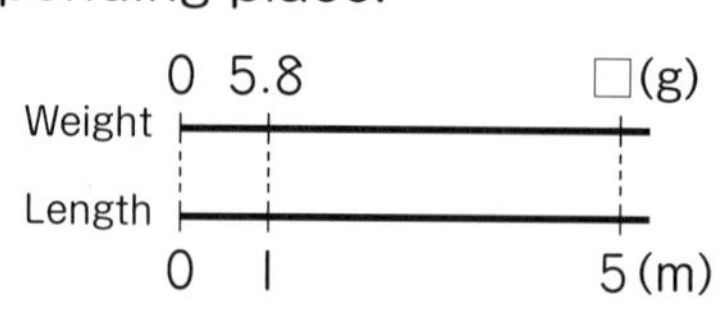

How to create a 4-square relationship table ▷

(1) Draw a table of 4 cells.

(2) Since the weight of 1 m is 5.8 g, write "1 m" and "5.8 g" in the left column cells.

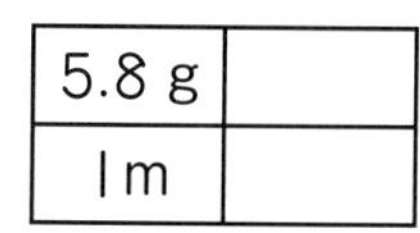

5.8 g	
1 m	

(3) Since we do not know the weight of 5 m, let □ g be the weight of 5 m, and write "5 m" and "□ g" in the right column cells.

5.8 g	□ g
1 m	5 m

You can write the 4-square relationship table either in alignment with the diagram or in the order in which they appear in the problem text, as shown on the right.

1 m	5 m
5.8 g	□ g

I'll get the same unit of measure for the horizontal squares.

Sara

In mathematics, sometimes it is easier to understand a problem situation if it is represented by a diagram or a table. Let's look back at the different diagrams we have seen so far.

When you don't know the number for each

A wire of 6 m has a weight of 25.2 g. What is the weight of 1 meter of this wire?

Think of it as a diagram...

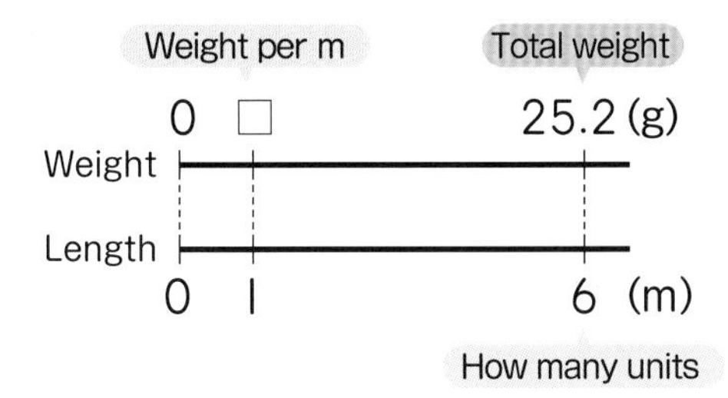

Total weight ÷ How many units = Weight per m

Therefore,

$25.2 \div 6 = 4.2$ Answer: 4.2 g

Think of it as a table...

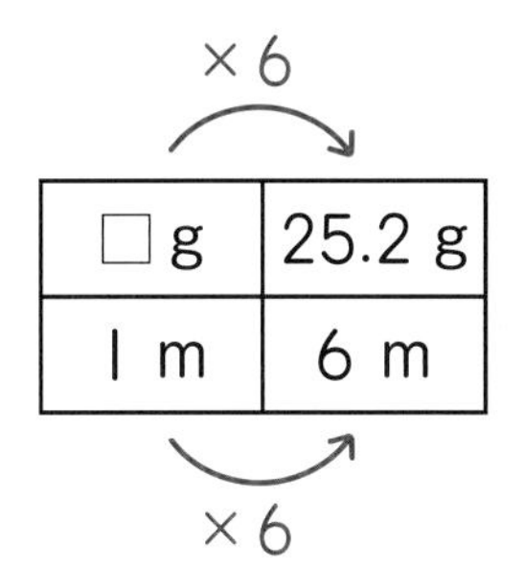

$\square \times 6 = 25.2$

$25.2 \div 6 = 4.2$ Answer: 4.2 g

When you don't know the number of units

A wire weighs 3 g per meter. You cut off several meters of this wire and weighed it, and it weighed 46.5 g. How many meters did you cut off?

Think of it as a diagram....

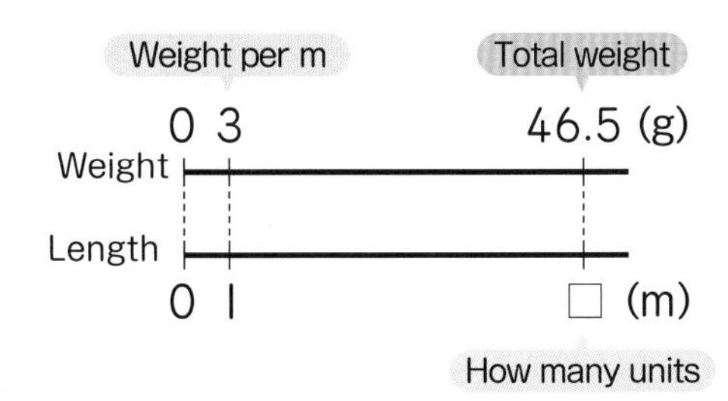

Total weight ÷ Weight per m = How many units

Therefore,

$46.5 \div 3 = 15.5$ Answer: 15.5 m

Think of it as a table...

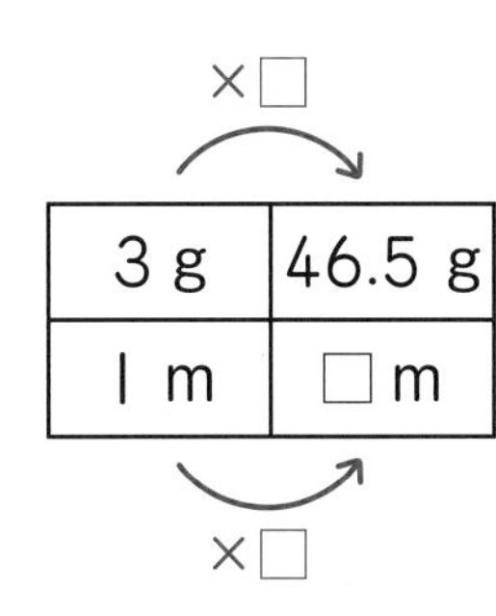

$3 \times \square = 46.5$

$46.5 \div 3 = 15.5$ Answer: 15.5 m

You can think of it as a figure or as a table, whichever is easier to understand.

Find the ?

How much does the ribbon cost?

Have you heard of the Citrus Ribbon Project? It is a project that aims to create a caring and comfortable society where all people can say "I' m home" and "welcome home. The three rings of the Citrus Ribbon represent "Region," "Home," and "School."

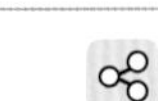

1

Let's make citrus ribbon.

If we want to make one citrus ribbon, we need 60 cm of ribbon.

Since we are four, we will need 240 cm.

Let's go buy enough ribbon for what we need.

2

I m costs 80 yen

240 cm is 2.4 m so ...

How much would it be

3

\ Want to think /

What operation should we do to find the price for 2.4 meters?

Multiplication of Decimal Numbers

Let's think about how to calculate and use the rules of operations.

1 Operations of whole number x decimal number

1 You buy a ribbon that costs 80 yen per meter. How much would you pay for 2.4 m of ribbon?

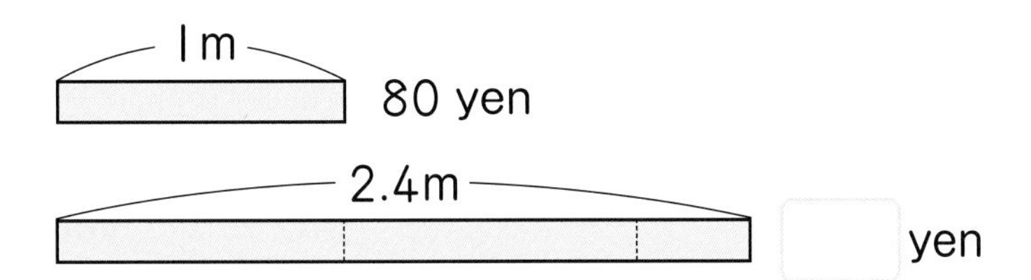

1 Let's consider a math equation to find the total price by using a diagram and a table.

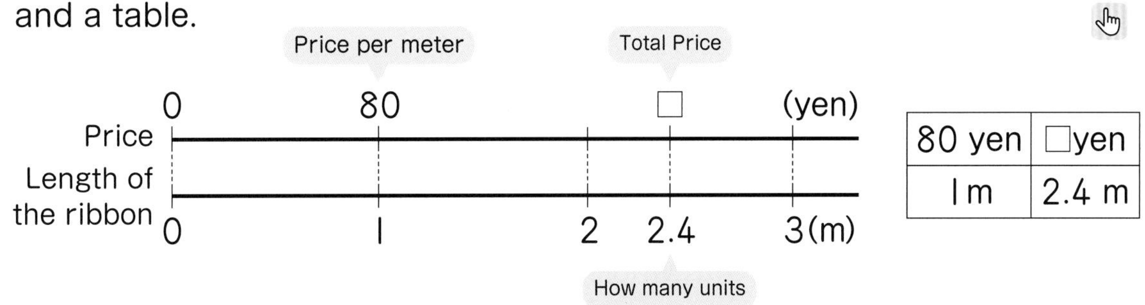

80 yen	□yen
1 m	2.4 m

If it is 2 m, the operation is 80 × 2, if it is 3 m, the operation is 80 × 3...

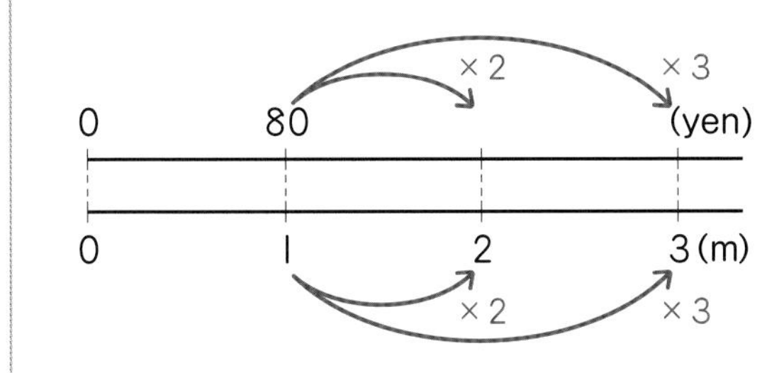

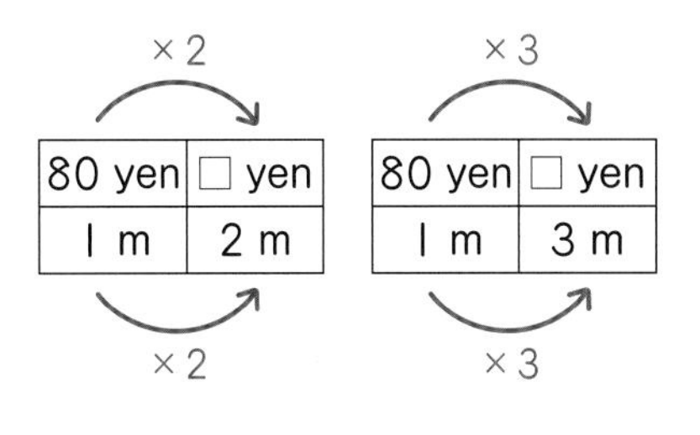

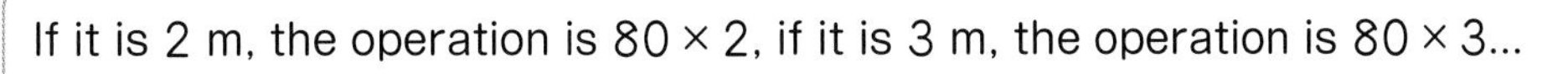

If the price is 80 yen per meter, then for 2.4 meters, the total cost is....

If the price is proportional to the length of the ribbon...

② Let's explain the ideas of the following children.

Akari's idea

Price per meter × Length = Total price

then,

	Price per meter		Length		Total price
When the length of the ribbon is 2 m	80	×	2	=	160
3 m	80	×	3	=	240
2.4 m	80	×	□	=	□

Way to see and think

When the "how many units" is a decimal number, can we do the same as with whole numbers?

Yu's idea

When the length increases by 2.4 times, the price also increases by 2.4 times.

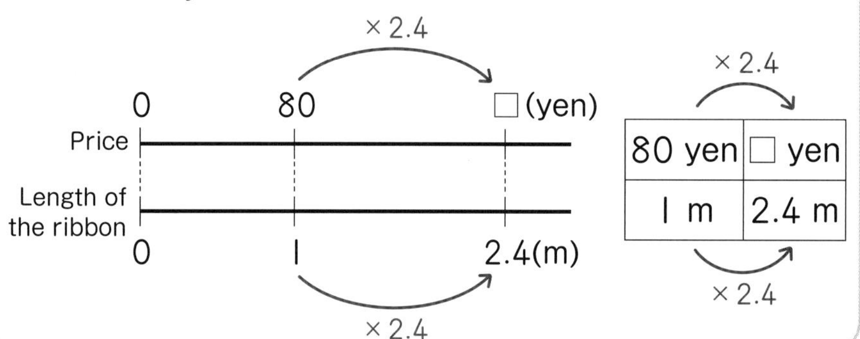

×2.4

80 yen	□ yen
1 m	2.4 m

×2.4

Way to see and think

If the length of the ribbon increases by 2.4 times, does the price increase by 2.4 times as well?

Summary

If the "how many units" is a decimal number, like the length of the ribbon, the total measurement can be calculated with a multiplication in the same way as with whole numbers.

③ What would the total price be?

2 m is 160 yen and 3 m is 240 yen, so it would be somewhere in between...

0	80	160	□	240 (yen)
0	1	2	2.4	3 (m)

Sara

Since 2.4 m is about half of 5 m, it would be about half of 400 yen for 5 m.

Haruto

④ Let's think about how to calculate 80 × 2.4.

? What kind of calculation should we do to multiply with decimals?

2 Let's compare how the following two children calculate 80×2.4.

\ Want to think /

Purpose Think about how to calculate whole numbers x decimals.

Sara's idea

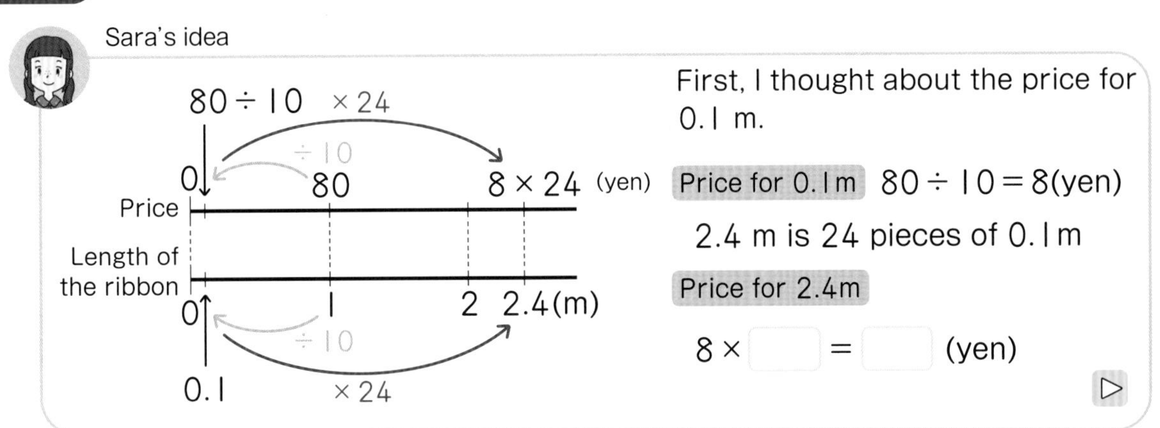

First, I thought about the price for 0.1 m.

Price for 0.1m $80 \div 10 = 8$(yen)

2.4 m is 24 pieces of 0.1m

Price for 2.4m

$8 \times \square = \square$ (yen)

Haruto's idea

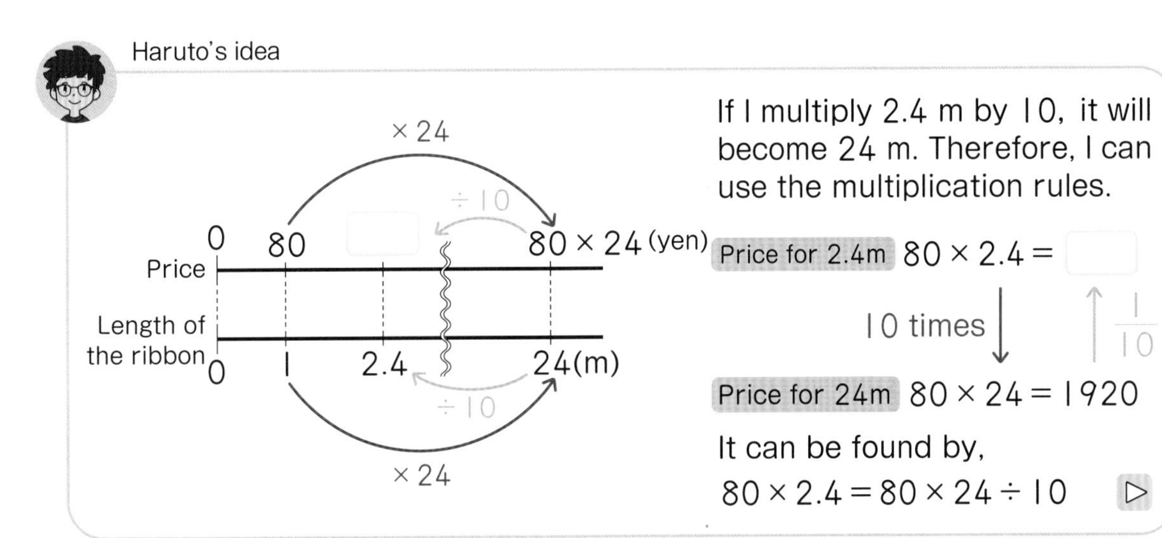

If I multiply 2.4 m by 10, it will become 24 m. Therefore, I can use the multiplication rules.

Price for 2.4m $80 \times 2.4 = \square$

10 times ↓ ↑ $\frac{1}{10}$

Price for 24m $80 \times 24 = 1920$

It can be found by,

$80 \times 2.4 = 80 \times 24 \div 10$

1 What is the price for 2.3 m of ribbon, if the price per meter is 90 yen? Let's find out using Sara's or Haruto's idea from above.

Summary

For the calculation of whole number × decimal number, if the decimal number is changed to a whole number then the answer can be found.

I could calculate decimals × whole numbers by vertical form, but can I also calculate whole numbers x decimals by vertical form?

3 Let's think about how to calculate 80×2.4 in vertical form.

		8	0
	×	2.	4

Akari: Can I use the method for decimal x whole numbers?

Yu: The position of the decimal is the same as in decimal x whole number?

Want to think

Purpose **What should we do to calculate whole number x decimal number in vertical form?**

Who had the same idea in exercise 2 on page 97?

1 Let's explain how to calculate in vertical form.

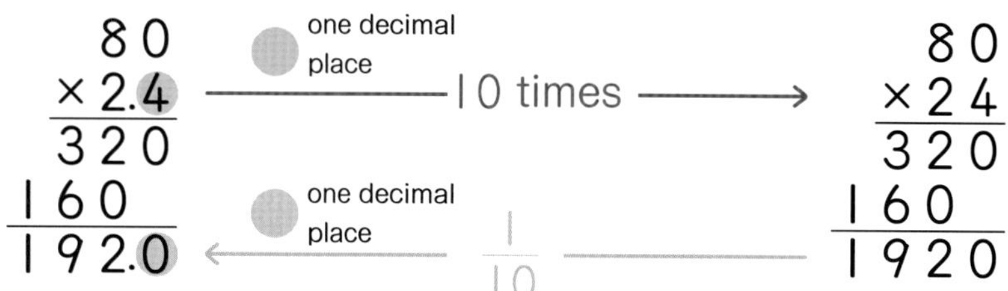

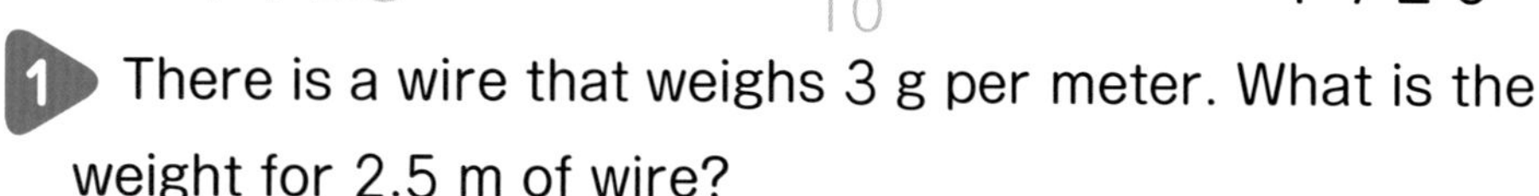

1 There is a wire that weighs 3 g per meter. What is the weight for 2.5 m of wire?

① Let's write a math expression.

② Let's calculate in vertical form.

		3
×	2.	5

Summary Way to see and think

Whole number × decimal number calculation is done in the same way as whole numbers calculations, assuming there is no decimal point. The number of digits after the decimal point of the product is the same as the number of decimals places in the decimal number.

2 Let's solve the following calculations in vertical form.

① 60×4.7　② 50×3.9　③ 7×1.6

④ 6×2.7　⑤ 24×3.3　⑥ 13×2.8

? Since we could do decimal x whole number and whole number x decimal, can we do decimal x decimal?

2 Operation of decimanl number x decimal number

1 I dL of paint was used to paint 2.1 m^2 of a wall.
What area, in m^2, can be painted with 2.3 dL of paint?

① Using a diagram and a table, let's think of a math equation to find the painted area.

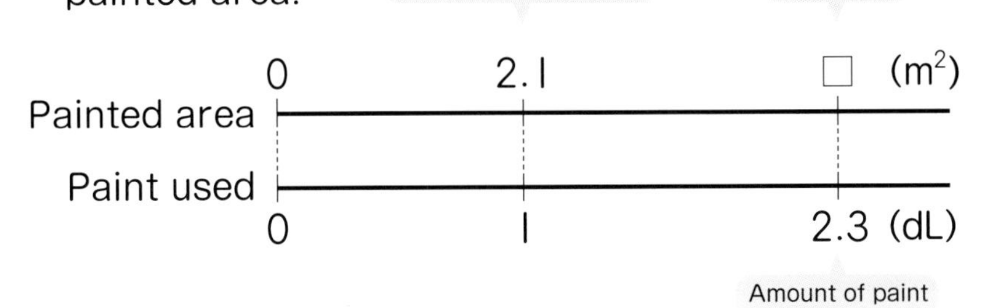

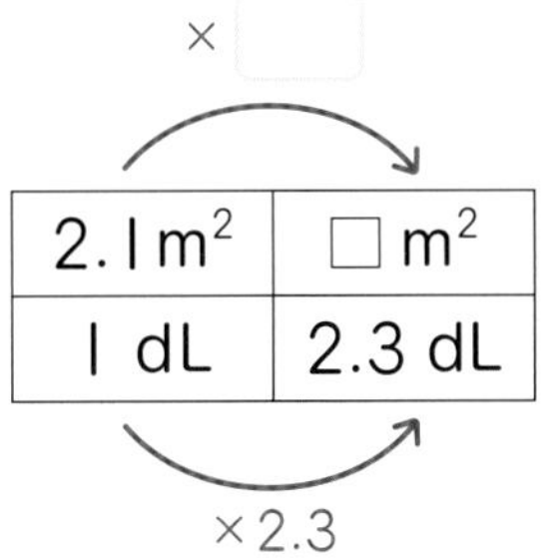

2.1 m^2	□ m^2
I dL	2.3 dL

② Let's write a math expression.

③ Let's explain the calculation methods of the following two children.

Akari's idea

Since I can calculate decimal number x whole number, by using the rules of multiplication.

$2.1 \times 2.3 = \square$

10 times ↓ ↑ $\frac{1}{10}$

becomes $2.1 \times 23 = \square$

Yu's idea

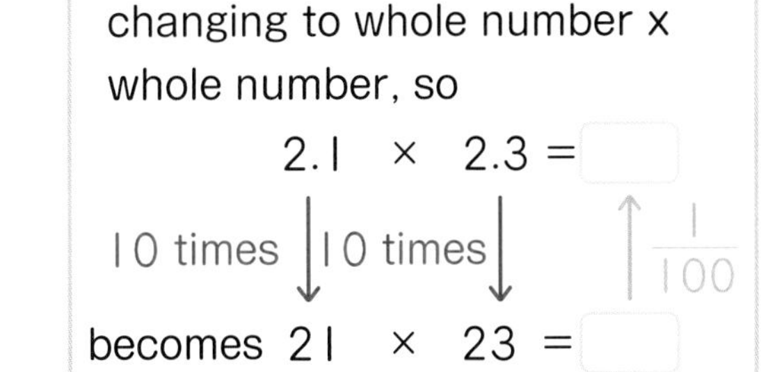

I can calculate easily by changing to whole number x whole number, so
2.1 × 2.3 =
10 times 10 times
1/100
becomes 21 × 23 =

Is it possible to use the vertical form for whole number x decimal number?

Is it possible to calculate decimal x decimal in vertical form?

Want to think

Is the calculation of decimal number × decimal number in vertical form the same as before?

④ Let's explain how to calculate 2.1 × 2.3 in vertical form.

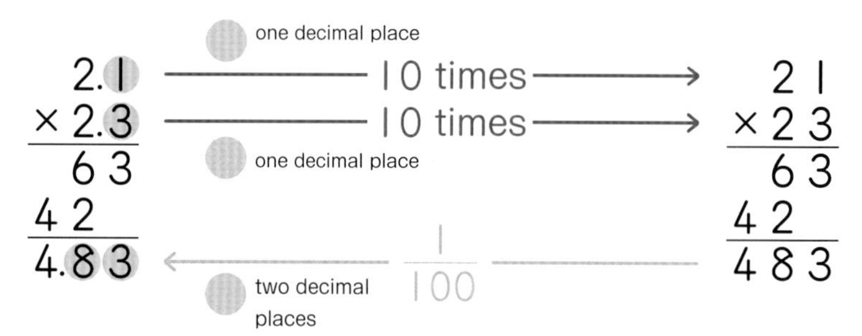

$$\begin{array}{r} 2.1 \\ \times\ 2.3 \\ \hline 63 \\ 42\ \\ \hline 4.83 \end{array} \qquad \begin{array}{r} 21 \\ \times\ 23 \\ \hline 63 \\ 42\ \\ \hline 483 \end{array}$$

1 Let's think about how to calculate 5.26×4.8 in vertical form.

two decimal places — 100 times →

one decimal place — 10 times →

← $\frac{1}{1000}$ — three decimal places

$$\begin{array}{r} 5.26 \\ \times\ 4.8 \\ \hline 4208 \\ 2104\ \\ \hline 25.248 \end{array} \qquad \begin{array}{r} 526 \\ \times\ 48 \\ \hline 4208 \\ 2104\ \\ \hline 25248 \end{array}$$

		5	. 2	6
	×		4	. 8

Summary — Way to see and think

For the previous calculations, decimals x decimals can be solve ignoring the decimal point, and solve as if they were whole numbers. But for give the answer at the end, the position of the decimal point in the product should be considered.

? When we add a decimal point to a product, what if the place below the decimal point is 0?

Multiplication algorithm of decimal numbers in vertical form ▷

① Calculate in the same way as the calculation of whole numbers, assuming that there is no decimal point.

② Place the decimal point of the product such that the number of decimal places is the same as the sum of the decimal places of the numbers multiplied.

$$\begin{array}{r} 2.1 \\ \times\ 2.3 \\ \hline 63 \\ 42\ \\ \hline 4.83 \end{array}$$

2.① ···· 1 decimal place
× 2.③ ···· 1 decimal place
4.⑧③ ···· ② decimal places

$$\begin{array}{r} 5.26 \\ \times\ 4.8 \\ \hline 4208 \\ 2104\ \\ \hline 25.248 \end{array}$$

5.②⑥ ···· 2 decimal places
× 4.⑧ ···· 1 decimal place
25.②④⑧ ···· ③ decimal places

2 Let's think about how to calculate 4.36×7.5.

\ Want to think /

Purpose Consider various decimal multiplications.

Akari

① What is the approximate answer?

② Let's think about it in vertical form.

		4.	3	6
	×		7.	5
	2	1	8	0
3	0	5	2	
3	2.	7	~~0~~	~~0~~

4.36 —— ☐ times → 436

7.5 —— ☐ times → 75

32.700 ← ☐ —— 32700

		4	3	6
	×		7	5
	2	1	8	0
3	0	5	2	
3	2	7	0	0

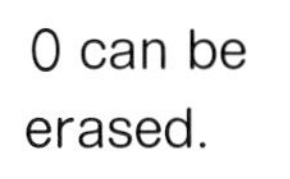

1 In the following calculations, let's place the decimal point of the product.

①
```
   5.6
×  1.3
  168
 224
 2408
```

②
```
  3.27
×  1.2
   654
  327
  3924
```

③
```
  1.48
×  2.5
   740
  296
  3700
```

2 Let's calculate the following in vertical form.

① 0.2×1.6

	0.	2
×	1.	6

② 0.4×0.35

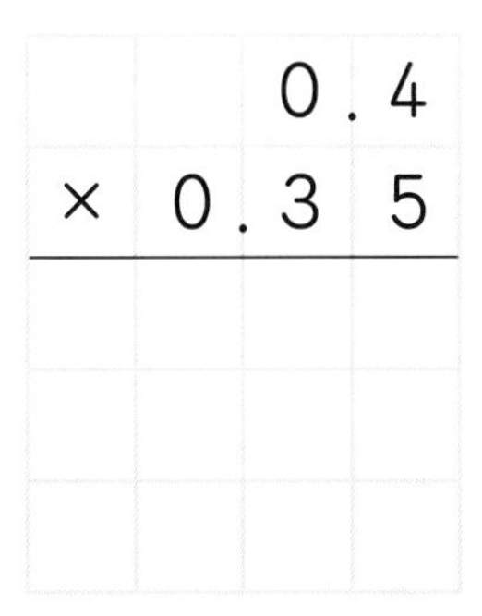

		0.	4
×	0.	3	5

Summary If you are careful about the position of the decimal point and handling the zeros, you can do it just like multiplying whole numbers.

Yu

3 Let's calculate the following in vertical form.

① 1.2×2.4 ② 6.4×3.5 ③ 3.14×2.6

④ 1.4×4.87 ⑤ 8.2×2.25 ⑥ 0.3×1.7

⑦ 0.5×2.3 ⑧ 0.43×2.1 ⑨ 1.15×0.6

? Multiplying by a number less than 1, the size of the product is...?

3

A metal bar weighs 3.2 kg per meter. Let's find the weight of two metal bars with a length of 1.2 m and 0.8 m.

The weight of the bar that is 1.2 m is 1.2 times the weight of a bar that is 1 m.

0.8 m is shorter than 1 m.

Want to explore

? Purpose What is the size of the product when the multiplier is a decimal number larger than 1 and smaller than 1?

1. Let's write a math expression to find the weight of the metal bar that is 1.2 m.

2. Let's write a math expression to find the weight of the metal bar that is 0.8 m.

3. Let's explain about the size of the product and size of the multiplicand using the following number line.

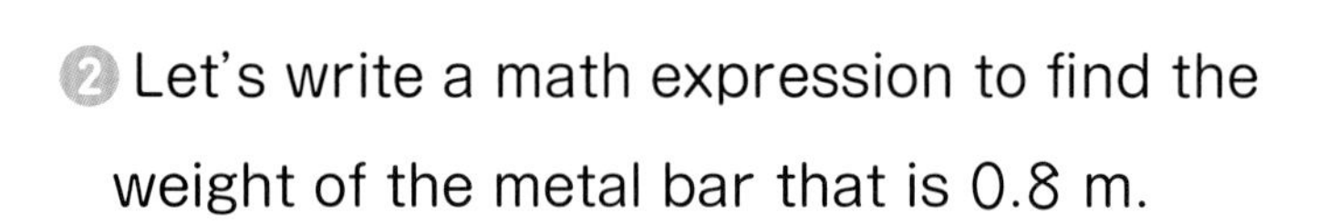

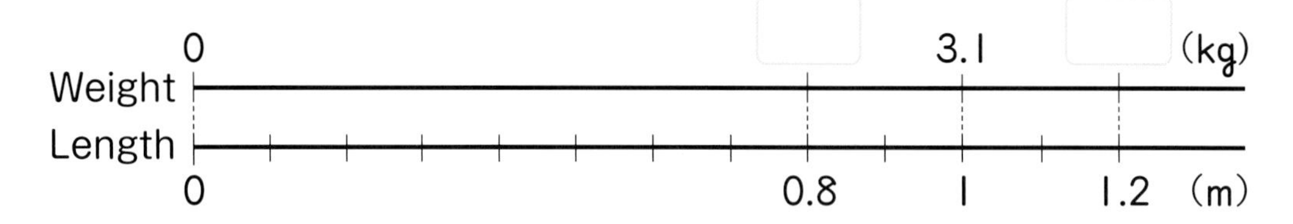

1 A cooking oils weighs 0.9 kg per liter. Let's find the weight of 0.4 L of this cooking oil.

① Let's write a math expression.

② Let's calculate in vertical form.

Since 0.4 is smaller than 1, then the weight of 0.4 L is less than 0.9 kg.

When the multiplier is a decimal number larger than 1, the product is larger than the multiplicand.
When the multiplier is a decimal number smaller than 1, the product is smaller than the multiplicand.
When the multiplier is 1, the product is the same as the multiplicand.

2 Let's place the decimal point of the products and compare products and multiplicands.

① $\begin{array}{r} 25 \\ \times\ \ 6 \\ \hline 150 \end{array}$ $\begin{array}{r} 25 \\ \times 0.6 \\ \hline 150 \end{array}$

② $\begin{array}{r} 2.5 \\ \times\ \ 6 \\ \hline 150 \end{array}$ $\begin{array}{r} 0.25 \\ \times\ 0.6 \\ \hline 150 \end{array}$

③ $\begin{array}{r} 1.6 \\ \times 0.24 \\ \hline 384 \end{array}$ $\begin{array}{r} 1.6 \\ \times 2.4 \\ \hline 384 \end{array}$

④ $\begin{array}{r} 0.07 \\ \times\ 0.2 \\ \hline 14 \end{array}$ $\begin{array}{r} 0.07 \\ \times\ \ 2 \\ \hline 14 \end{array}$

3 Using six of the following 8 cards, let's make an equation that can be calculated in vertical form as shown on the right.

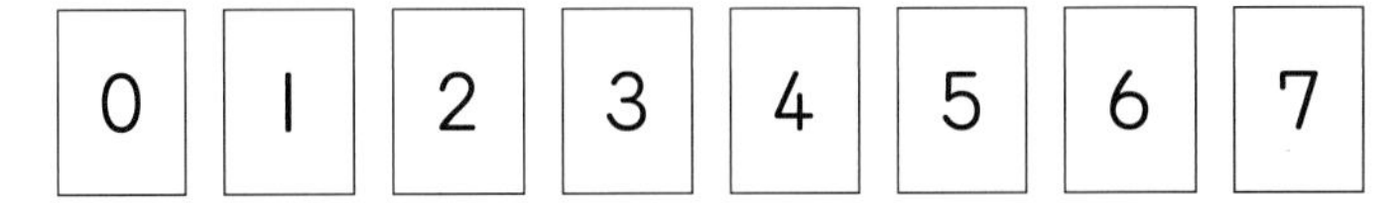

In addition, you cannot place 0 in ☐ of ⓐ or ⓑ. Each card can only used once.

Let's answer the following questions.

① In the case that the multiplicand is 13, let's find the numbers that apply to the above calculation.

② The multiplier will never be 0.1 nor 0.5 for any number you use as the multiplicand. Let's explain the reason.

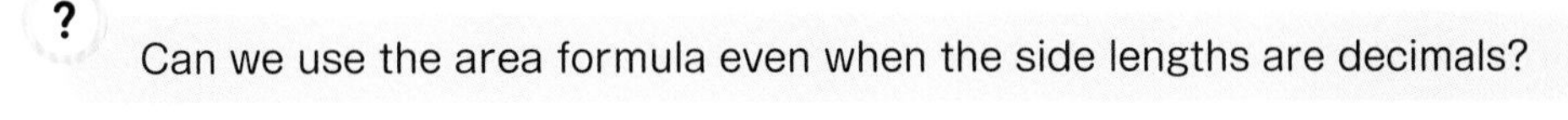

?
Can we use the area formula even when the side lengths are decimals?

4 What is the area, in m^2, of a rectangular flower bed that has a length of 2.4 m and a width of 3.1 m?

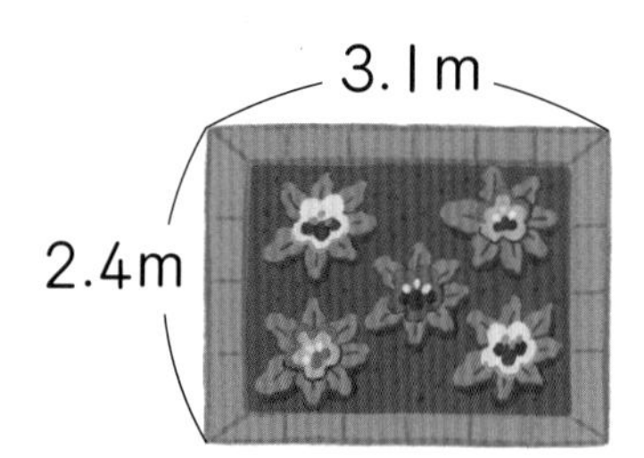

Can I think about it using the diagram?

Can I use the formula for the area?

\ Want to know /

? Purpose **Even when the length and width are decimal numbers, can we use the formula for area?**

Yu's idea

$1\,m^2$ is divided into 10 squares for each side, therefore $0.01\,m^2$ represents $\frac{1}{100}$ of $1\,m^2$. I thought how many of these parts is each side of the flower bed.

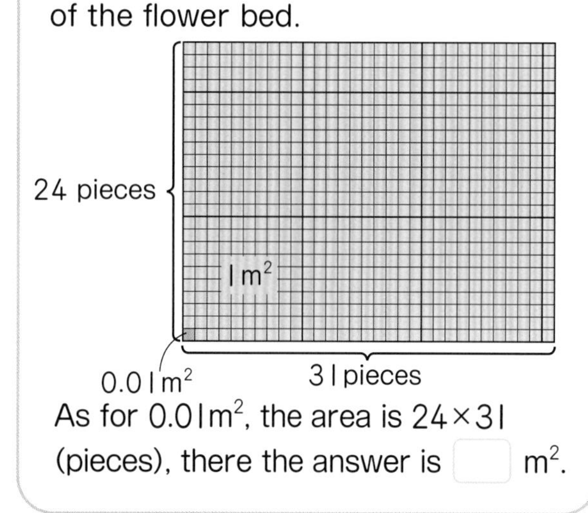

As for $0.01\,m^2$, the area is 24×31 (pieces), there the answer is ☐ m^2.

Sara's idea

If I apply the formula for area, 2.4×3.1 and consider solve it in the vertical form.

	2	. 4
×	3	. 1

Then the area is ☐ m^2.

! Summary

The area can be found using the formula even if the lengths of the sides are decimal numbers.

1 What is the area, in m^2, of a rectangular flower bed that has a length of 0.6 m and a width of 2.5 m?

? Do the rules of operations for whole numbers apply also to decimals?

3 Rules of operations

1 When doing calculation with whole numbers, the following rules of operations are valid. Let's explore if the rules are also valid with decimal numbers.

ⓐ ■ × ▲ = ▲ × ■ (Rule of Commutativity)

ⓑ (■ × ▲) × ● = ■ × (▲ × ●) (Rule of Associativity)

ⓒ (■ + ▲) × ● = ■ × ● + ▲ × ●
ⓓ (■ − ▲) × ● = ■ × ● − ▲ × ●
} (Rule of Distribution)

Akari: When two whole numbers are multiplied, the product is the same even if the multiplicand and the multiplier are interchanged.

Haruto: I wonder if the same can be said for decimals.

\ Want to explore /

? **Purpose** Are the rules of operations also valid for decimal numbers?

① Let's explain if the rule ⓐ is valid.
Use the diagram on the right.

$3.6 \times 2.4 =$ ☐

$2.4 \times 3.6 =$ ☐

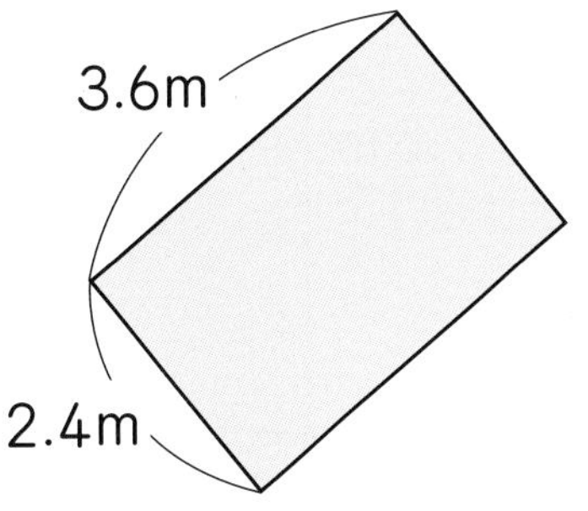

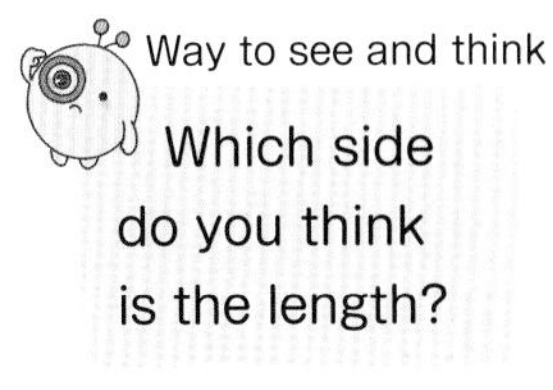

Way to see and think
Which side do you think is the length?

② Let's explain if the rule ⓑ is valid.
Use the diagram on the right.

$(3.6 \times 2.5) \times 4 =$ ☐

$3.6 \times (2.5 \times 4) =$ ☐

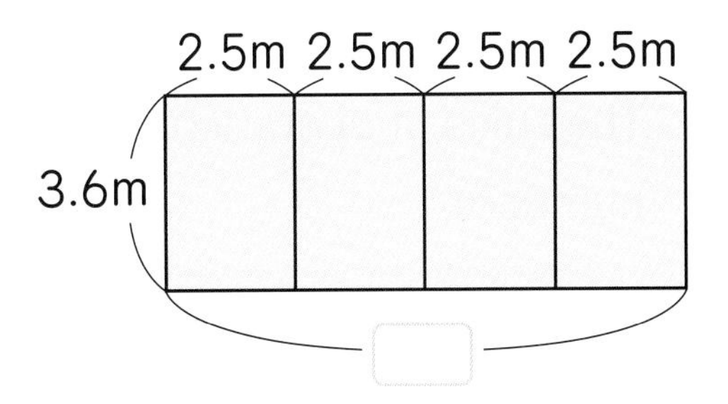

Way to see and think
The equation changes depending on whether you see 4 small rectangles or only 1 large rectangle.

3. Let's explain if the rule ⓒ is valid. Use the diagram below.

$(1+0.4)\times 3=$ ☐

$1\times 3+0.4\times 3=$ ☐

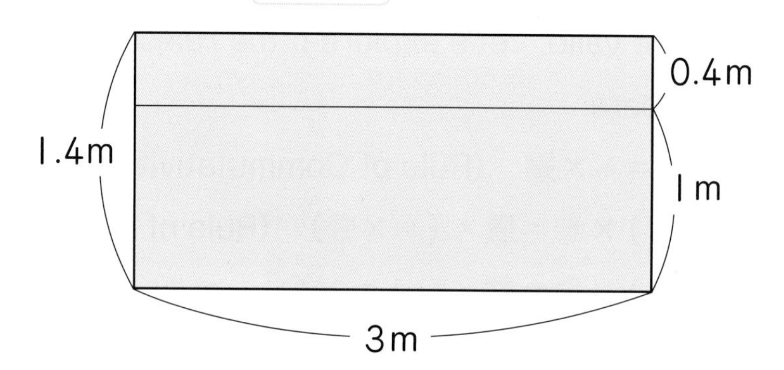

4. Let's explain if the rule ⓓ is valid. Use the diagram below.

$(2-0.2)\times 3=$ ☐

$2\times 3-0.2\times 3=$ ☐

0.2m
1.8m
2m
3m

Summary — Way to see and think

The rules of operations are valid for whole numbers and decimal numbers.

1 Let's calculate by using the rules of operations. Let's also write the calculation process.

① $6.9\times 4\times 2.5$

② $0.5\times 4.3\times 4$

③ $3.8\times 4.8+3.8\times 5.2$

④ $1.3\times 12.9-1.3\times 2.9$

CAN What can you do?

☐ We understand how to multiply decimal numbers. → p.99

1 Let's summarize how to multiply decimal numbers.

To calculate 2.3 × 1.6, first multiply 2.3 by ☐ and multiply 1.6 by ☐ .

Then calculate ☐ × ☐ . Finally, the answer 368 is multiplied by ☐ .

Therefore, 2.3 × 1.6 = ☐

☐ We can calculate whole number x decimal number and decimal number x decimal number in vertical form. → pp.98 ~ 101

2 Let's calculate the following in vertical form.

① 50 × 4.3	② 6 × 1.8	③ 26 × 3.2	④ 4 × 1.9
⑤ 31 × 5.2	⑥ 62 × 0.7	⑦ 0.6 × 0.8	⑧ 3.3 × 0.9
⑨ 1.5 × 3.4	⑩ 0.3 × 0.25	⑪ 1.26 × 2.3	⑫ 1.5 × 4.36
⑬ 27 × 3.4	⑭ 3.2 × 1.8	⑮ 0.4 × 0.6	⑯ 7.6 × 0.5
⑰ 2.87 × 4.3	⑱ 0.07 × 0.8	⑲ 4.5 × 0.06	⑳ 0.2 × 0.5

☐ We understand the relationship between the multiplier and the product. → pp.102 ~ 103

3 A wire weighs 4.5 g per meter. Let's find the weight of it if the length is 8.6 m and 0.8 m.

☐ We can represent in a math expression using decimal numbers and find the answer. → pp.104

4 Let's find the area of the rectangle shown on the right.

3.4m

1.2m

☐ We can calculate using the rules of operations. → pp.105 ~ 106

5 Let's calculate by using the rules of operations. Let's also write the calculation process.

① 7.4 × 4 × 2.5 ② 4.6 × 1.9 + 5.4 × 1.9 ③ 6.8 × 0.5 − 2.8 × 0.5

Supplementary problems → p.161

Which "Way to See and Think Monsters" did you find in " 7 Multiplication of Decimal Numbers"?

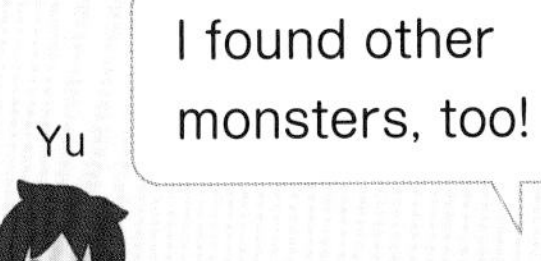

When I was thinking about how to calculate multiplication of decimal numbers, I found "Same Way."
Yu

I found other monsters, too!

Utilize Usefulness and Efficiency of Learning

1 Let's find the mistake in the following vertical form and write the correct answer inside the ( ).

①

```
     4.3
  × 3.1 4
  -------
     1 7 2
    4 3
  -------
  0.6 0 2
```
( )

②

```
     0.9 5
  ×    3.4
  --------
     3 8 0
   2 8 5
  --------
   3 2.3 0̸
```
( )

③

```
      6.4
  ×   2.5
  -------
    3 2 0
  1 2 8
  -------
  1.6 0̸ 0̸
```
( )

2 Instead of multiplying the number by 2.5, a friend added 2.5 to the number and got an answer of 12.3 by mistake. What should have been the answer to the original problem?

3 In the following calculations, let's fill in each □ with an equal or inequality sign.

① 3.5×3.5 □ 3.5 ② 3.5×0.1 □ 3.5

③ 3.5×0.9 □ 3.5 ④ 3.5×1 □ 3.5

4 Using 4 of the following 6 cards, create various decimal x decimal expressions.

[2] [3] [5] [6] [7] [8] □.□ × □.□

① Create a math expression to get the product as a whole number.
Also, let's explain how you came up with the multiplication.

□.□ × □.□ □.□ × □.□ □.□ × □.□

□.□ × □.□ □.□ × □.□ □.□ × □.□

② Create a math expression to get the largest product possible. Also, write down how you thought of it. □.□ × □.□

With the Way to See and Think Monsters...

Let's Reflect!

Let's reflect on which monster you used while learning "7 Multiplication of Decimal Numbers."

Same Way

For the calculation of decimal numbers, if the decimal number is changed to a whole number then the answer could be found in the same way as in the case of whole numbers.

① How did you calculate 2.1×2.3?

$2.1 \times 2.3 = \square$

(10 times ↓, 10 times ↓, ↑ $\frac{1}{100}$)

$21 \times 23 = \square$

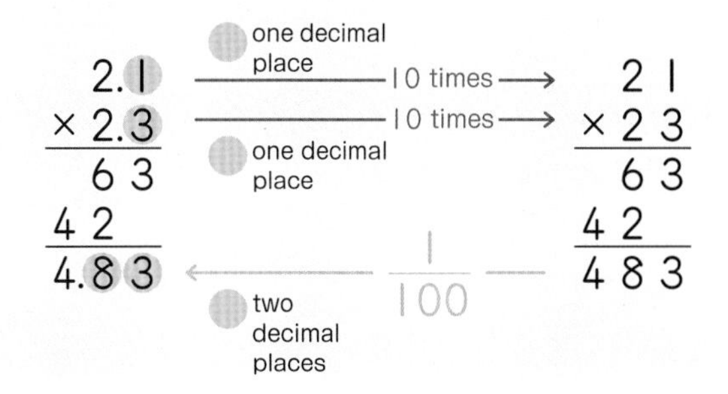

Multiply the multiplicand 2.1 and multiplier 2.3 by 10, respectively, in the same way as for whole numbers. The product is then multiplied by $\frac{1}{100}$.

Sara

In vertical form, I changed the multiplier and multiplicand to whole numbers, calculated them as whole numbers, and then added a decimal point at the end.

Akari

② Explain in a math expression that the calculation of 3.26×1.4 can be done using the calculation of 326×14.

$3.26 \times 1.4 = (0.01 \times \square) \times (0.1 \times \square)$

$= 0.01 \times 0.1 \times \square \times \square$

$= 0.001 \times \square$

$= 4.564$

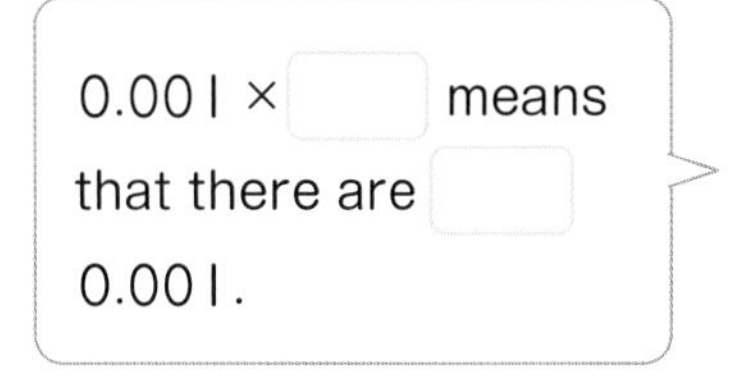

Haruto

? Solve the ?

Multiplication of decimal numbers could be done in the same way as we have learned by changing them to whole numbers.

Yu

Want to Connect

Can division using decimal numbers can be calculated in the same way as division using whole numbers?

Sara

Find the ?

Which one is a better deal?

▷

1

2 L
390 yen

1 L
200 yen

This juice costs 390 yen for 2 L, and 200 yen for 1L.

The 2L pack is a better deal since the price per liter is $390 \div 2 = 195$, which is 195 yen.

2

1.8 L
360 yen

1 L
210 yen

This juice costs 360 yen for 1.8 L, and 210 yen for 1L.

Which is cheaper per liter?

3

How about finding out the measure per unit quantity?

Since I divided by 2 for 2L, should I divide 360 by 1.8?

\ Want to think /

Purpose Can we divide by decimals?

Division of Decimal Numbers

Let's think about how to calculate.

1 Operation of whole number ÷ decimal number

1 If 1.8 L of juice costs 360 yen, what is the price per liter?

① Let's use a diagram and a table to think about how to write math expression to find the price.

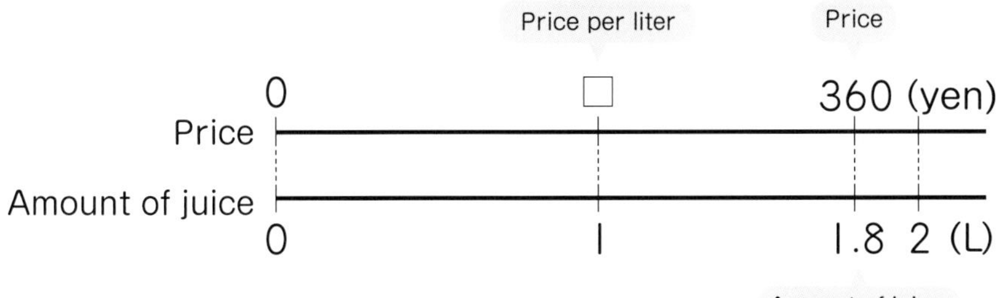

□ yen	360 yen
1 L	1.8 L

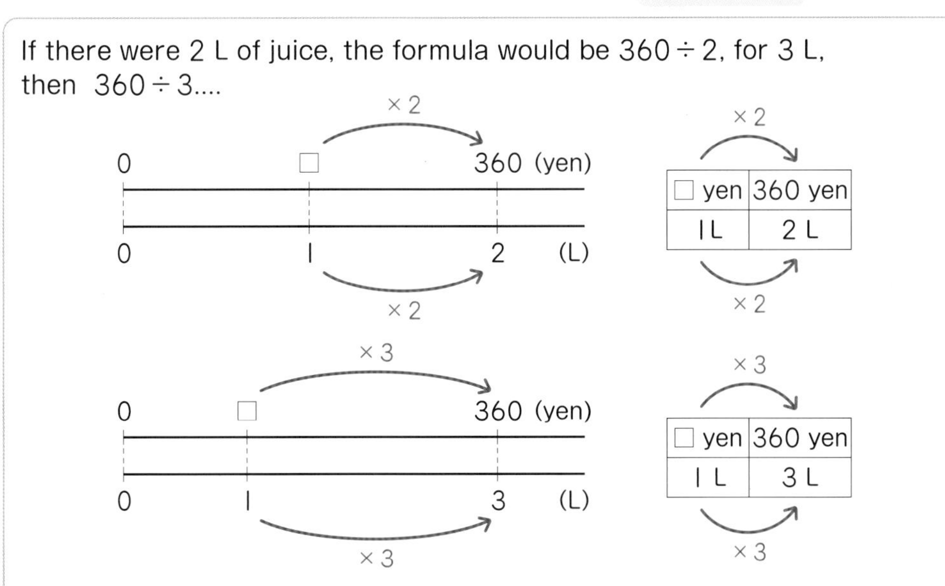

If there were 2 L of juice, the formula would be 360 ÷ 2, for 3 L, then 360 ÷ 3....

□ yen	360 yen
1 L	2 L

(×2)

□ yen	360 yen
1 L	3 L

(×3)

Haruto

Akari
Can we use the same idea of "times" as when we did in whole numbers?

② Let's explain the ideas of the following two children.

Haruto's idea

	Price		Amount of juice		Price per 1L
When the amount of juice is 2L	360	÷	2	=	180
3L	360	÷	3	=	120
1.8L	360	÷	1.8	=	□

Way to see and think

When the "how many units" is a decimal number, can you do the same as with whole numbers?

Akari's idea

If the amount of juice is 1.8 times, the price will be 1.8 times as well.

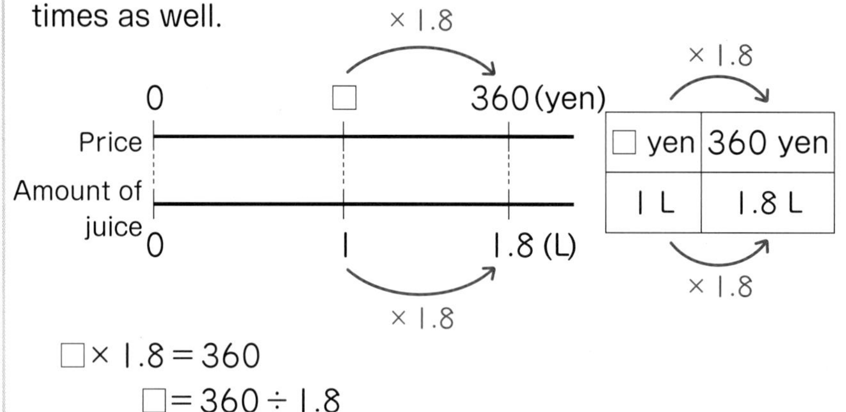

$\square \times 1.8 = 360$

$\square = 360 \div 1.8$

Way to see and think

Use the idea of number of times.

Summary

Even if the "how many units" is a decimal number, such as the amount of juice, the calculation of the measurement per unit will be a division in the same way as in the whole numbers.

③ About how much is the price?

④ Let's think about how to calculate $360 \div 1.8$.

Sara: Can we think about 1.8 as a whole number, just as we did with multiplication?

Haruto: I wonder if I can use the division rule.

? What kind of calculation should we do to divide with decimals?

2 Compare the following two calculations proposed to solve $360 \div 1.8$.

? **Purpose** ＼Want to think／ Think about how to calculate whole numbers ÷ decimals.

Sara's idea

1.8 L is equivalent to 18 times 0.1 L, **Price for 0.1 L** $360 \div 18 = 20$(yen)

Then the price of 1 L is 10 times 0.1 L, **Price for 1 L** $20 \times \square = \square$ (yen)

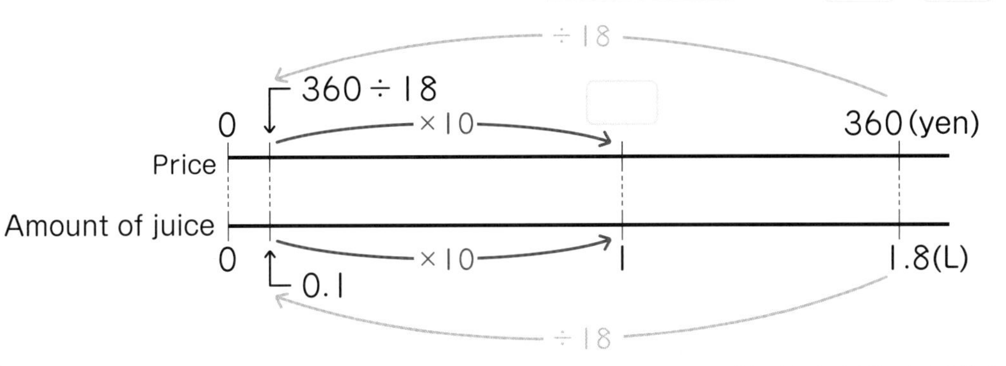

Haruto's idea

If you buy 10 times as much juice as 1.8 L, the price is also 10 times as much. Therefore, using the division rules:

When buying 1.8 L, 1 L cost $360 \div 1.8 = \square$ (yen)

10 times ↓ 10 times ↓

When buying 18 L, 1 L cost $3600 \div 18 = 200$ (yen)

1.5 L bottle of juice costs 240 yen. Let's find the price per liter using Sara's and Haruto's ideas above.

! **Summary**

To calculate a whole number divided by a decimal, you can find out the answer by converting the decimal into a whole number.

? I can calculate decimal ÷ whole number in vertical form, but can I also calculate whole numbers ÷ decimals in vertical form?

3 Let's think about how to calculate 360÷1.8 in vertical form

Can I use the idea of decimal ÷ whole number in vertical form?

Can I use the rules of division?

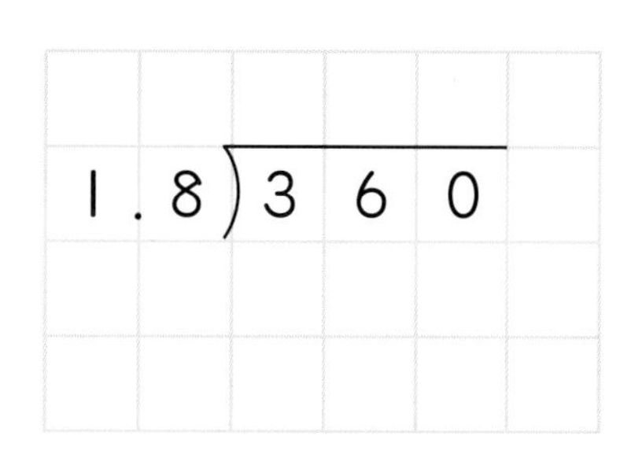

\ Want to think /

Purpose **What should we do to calculate whole number ÷ decimal number in vertical form?**

❶ Let's explain how to calculate the following in vertical form.

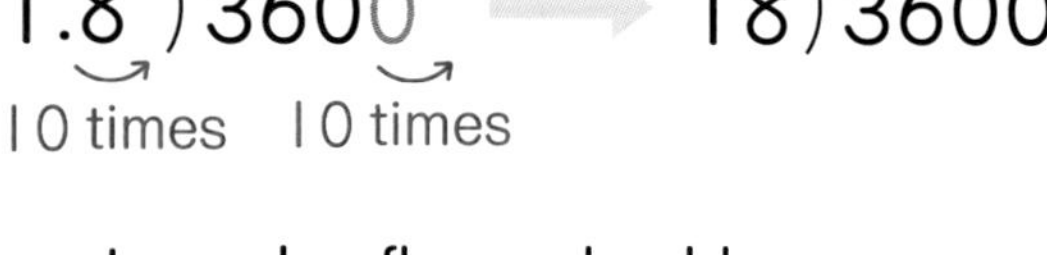

When you multiply by 10, the decimal point moves to the right.

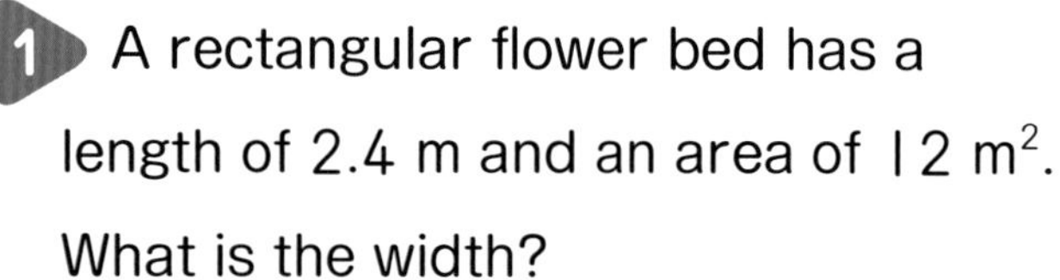

1 A rectangular flower bed has a length of 2.4 m and an area of 12 m^2. What is the width?

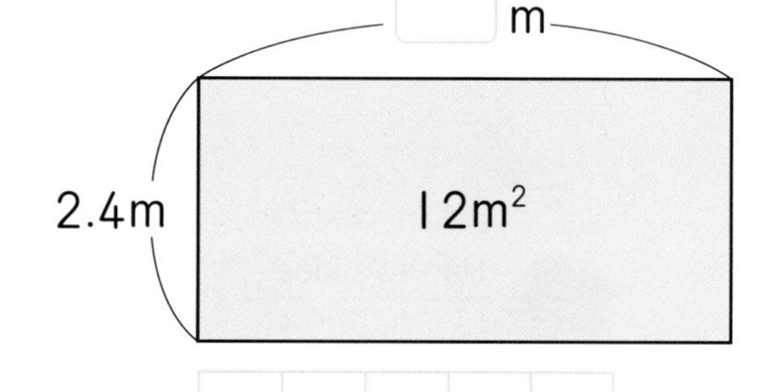

① Let's write a math expression.

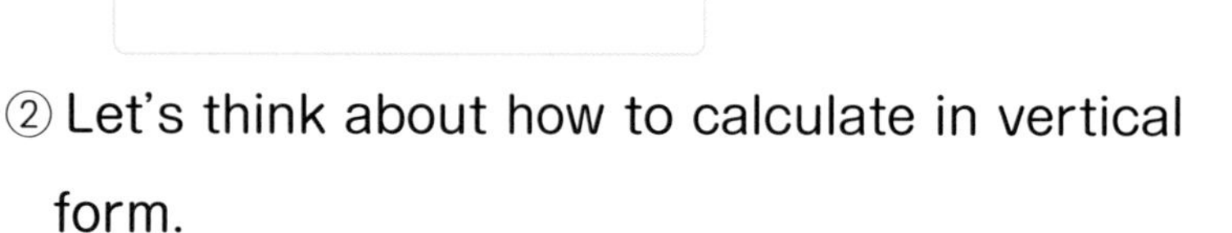

② Let's think about how to calculate in vertical form.

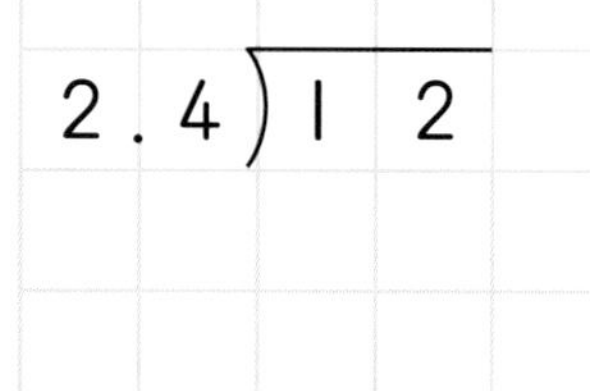

Summary

In division, the quotient does not change if the dividend and divisor are multiplied by the same number. When we divide a number by a decimal number, we can calculate it by changing the dividend and divisor to whole numbers by using the rules of division.

2 Let's calculate the following in vertical form.

① 9÷1.8 ② 91÷2.6 ③ 72÷4.8

? Since we can do decimal ÷ whole number and whole number ÷ decimal, do you think we can do decimal ÷ decimal?

2 Operation of decimal number ÷ decimal number

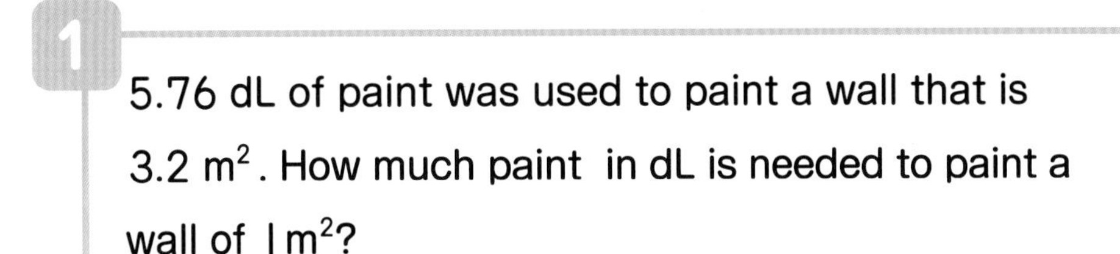

1 5.76 dL of paint was used to paint a wall that is 3.2 m^2. How much paint in dL is needed to paint a wall of 1 m^2?

① Let's think about a math expression to find the amount of paint, using a diagram and table.

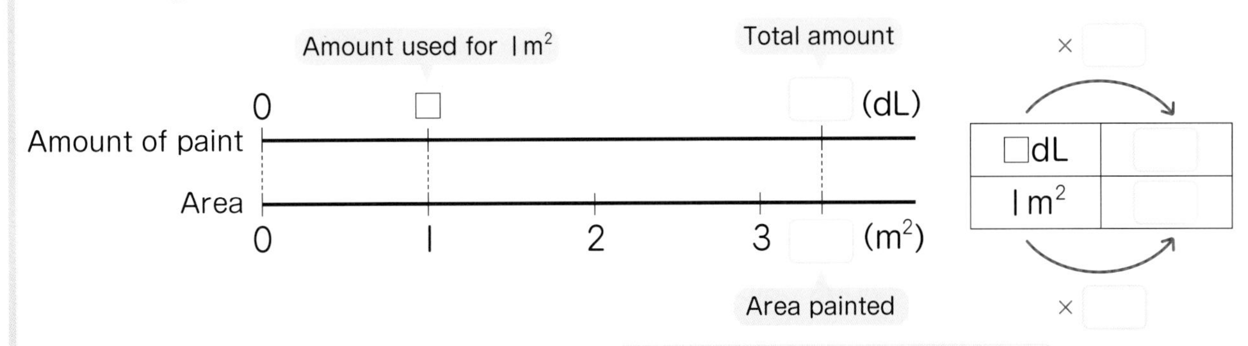

② Let's write a math expression.

③ Let's explain the calculation method of the following children.

Akari's idea

The paint needed for 0.1 m^2 is

$$5.76 \div 32 = 0.18 \text{ (dL)}$$

Therefore, the pain needed for 1 m^2 is 10 times that amount

$$0.18 \times 10 = \square \text{ (dL)}$$

Yu's idea

I will apply the rules of division to change the divisor to a whole number.

$$5.76 \div 3.2 = \square$$

10 times ↓ ↓ 10 times

$$57.6 \div 32 = \square$$

Sara: Can I use the vertical form to calculate whole number ÷ decimal number?

Haruto: Can I use the vertical form to calculate decimal ÷ decimal?

\ Want to think /

? Purpose Is the calculation of decimal number ÷ decimal number in vertical form the same as before?

④ Let's explain how to calculate $5.76 \div 3.2$ in vertical form.

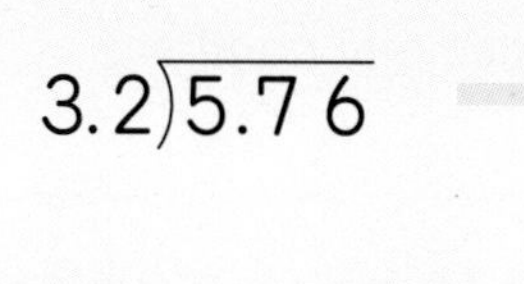

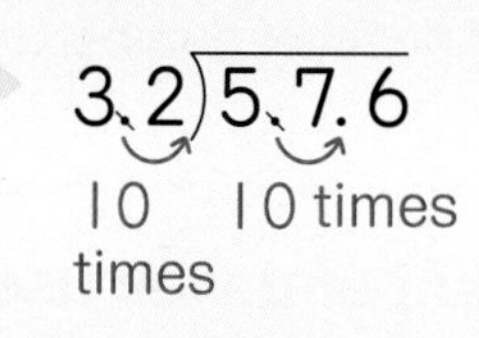


```
      1.8
3.2)5.7.6
    3 2
    2 5 6
    2 5 6
        0
```

1 A rectangular flower bed has an area of 8.4 m^2 and a width of 2.8 m. What is the length in meter?

① Let's write a math expression.

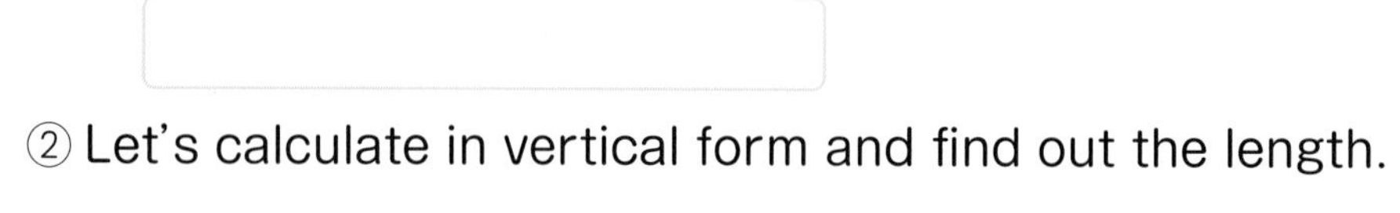

② Let's calculate in vertical form and find out the length.

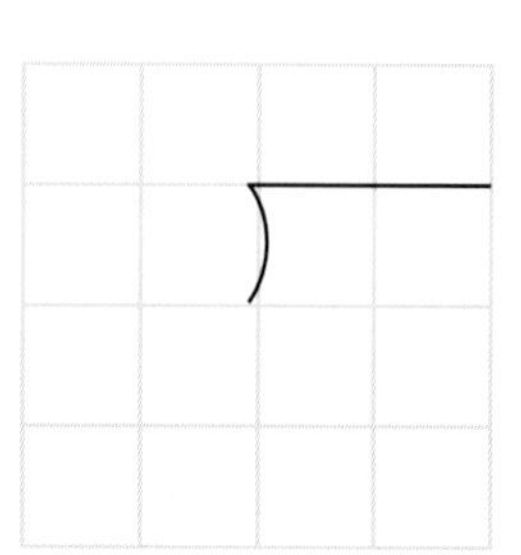

Summary

As with the previous calculations, decimal numbers ÷ decimal numbers can be written without the decimal point, as long as the position of the decimal point in the quotient is also considered by the change.

Way to see and think

2 Let's calculate the following in vertical form.

① $9.52 \div 3.4$	② $9.88 \div 2.6$	③ $7.05 \div 1.5$
④ $8.5 \div 1.7$	⑤ $7.6 \div 1.9$	⑥ $9.2 \div 2.3$

? In multiplication, when a number smaller than 1 is multiplied, the product is smaller than the number multiplied, but in division, when a number smaller than 1 is divided, is there any rule?

Division algorithm of decimal numbers in vertical form

① Multiply the divisor by 10, 100 or more to convert it into a whole number. Move the decimal point to the right accordingly.
② Multiply also the dividend by the same number as the divisor. And move the decimal point to the right accordingly.
③ The decimal point of the quotient is aligned at the same place as where the decimal point of the dividend has been moved to.
④ Then, calculate it as in division of whole numbers.

```
      1.8
3.2)5.7.6
    3 2
    2 5 6
    2 5 6
        0
```

2

There is a 1.2 m silver wire that weighs 19.2 g and a 0.8 m red wire that weighs 19.2 g. For each type, let's find the weight of a wire that is 1 m long.

\ Want to explore /

Purpose What is the size of the quotient when the divisor is a decimal number larger than 1 and smaller than 1?

① Let's find the weight per meter of each wire by writing a math expression.

【Silver wire】 □　　　【Red wire】 □

② Let's explain about the size of the quotient and the size of the dividend using the following number lines.

【Silver wire】

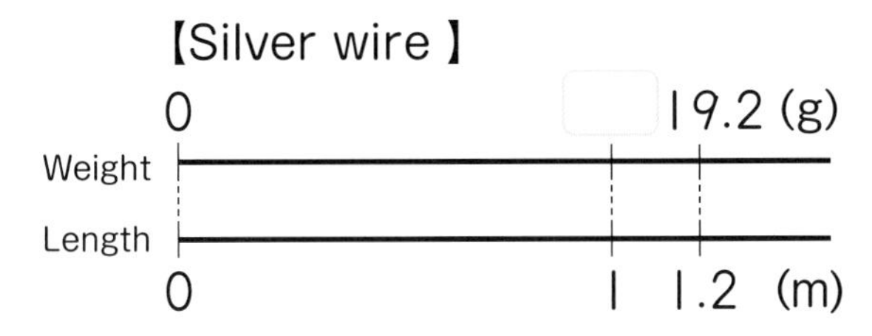

【Red wire】

0　19.2　□　(g)

Weight

Length

0　0.8　1　(m)

③ Let's calculate $19.2 \div$ □, by placing different numbers into □, except for 1.2 and 0.8. Let's compare the size with 19.2 as shown on the right.

$19.2 \div 0.8 = 24 > 19.2$

$19.2 \div 0.4 =$ □ □ 19.2

$19.2 \div 1 =$ □ □ 19.2

$19.2 \div 2 =$ □ □ 19.2

Summary

When a number is divided by a number larger than 1, the quotient becomes smaller than the dividend.
When a number is divided by a number smaller than 1, the quotient becomes larger than the dividend.
When a number is divided by 1, the quotient becomes the same as the dividend.

1 Does the quotient becomes larger than the dividend? Let's calculate and confirm.

① $49 \div 0.7$　② $1.5 \div 0.3$　③ $0.4 \div 0.2$

? The calculation of decimal number ÷ whole number is possible, but it is possible to do decimal ÷ decimal?

3

There is a 1.5 m metal bar that weighs 4.8 kg. What is the weight, in kg, of the bar if it is 1 m long?

① Let's think about it by using diagrams and tables.

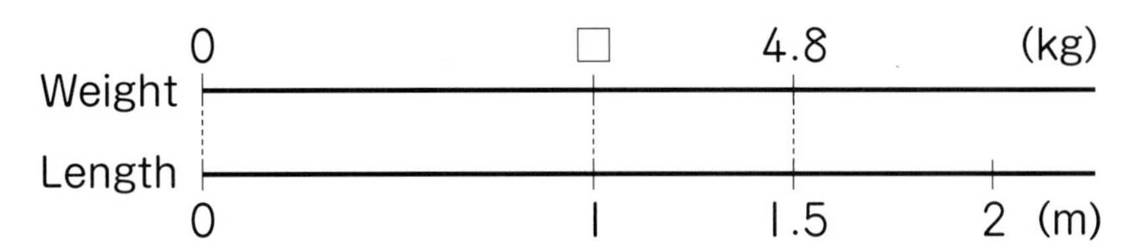

□kg	4.8 kg
1 m	1.5 m

② Let's write a math expression.

\ Want to think /

? Purpose **Can we divide decimals with decimals?**

③ Haruto thought about the calculation in vertical form as shown on the right. Let's think about how to continue with division.

④ To continue, Haruto thought of 48 as 48.0. Let's consider how to continue the calculation on the right to find the answer.

```
       3.
1.5)4.8.
    4 5
      3
```

```
       3.
1.5)4.8.0
    4 5 ↓
      3 0
```

! Summary

You can continue with the division of decimal numbers in vertical form, assuming there is a 0 in the smallest place.

1 Let's explain how to calculate 3.23 ÷ 3.8 in vertical form.

```
       0.8 5
3.8)3.2.3
     3 0 4
       1 9 0
       1 9 0
           0
```

Akari: Why is the quotient not written in the ones place?

2 Let's calculate the following in vertical form.

① 5.4 ÷ 1.5　② 3 ÷ 0.4　③ 36.9 ÷ 1.8　④ 3.06 ÷ 4.5

? If the number to be divided has two decimal places, can it be divided in the same way?

4 Let's think about how to calculate $7.85 \div 3.14$ in vertical form.

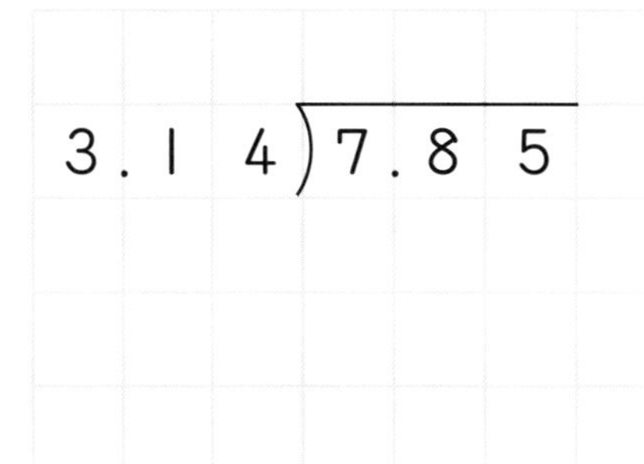

\ Want to think /

? **Purpose** Can we do the same division as before if we have two decimal places?

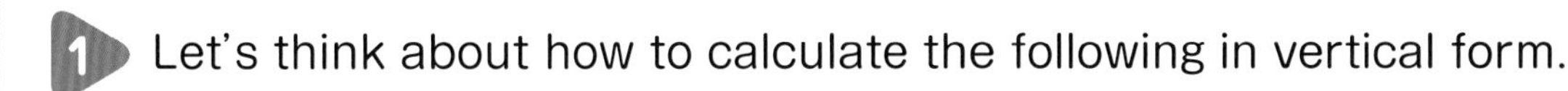

$$3.14\overline{)7.85} \rightarrow 3.14\overline{)7.85} \rightarrow 3.14\overline{)7.85}$$

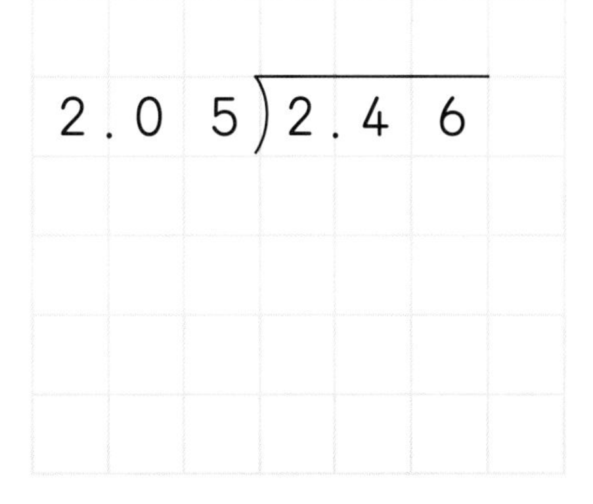

1 Let's think about how to calculate the following in vertical form.

① $4.64 \div 1.45$　　② $2.46 \div 2.05$

! **Summary** Way to see and think

The same division can be made if we have two decimal places, as long as the position of the decimal point is taken into account.

2 Let's calculate the following in vertical form.

① $1.75 \div 1.25$　　② $3.24 \div 1.35$　　③ $0.12 \div 0.48$

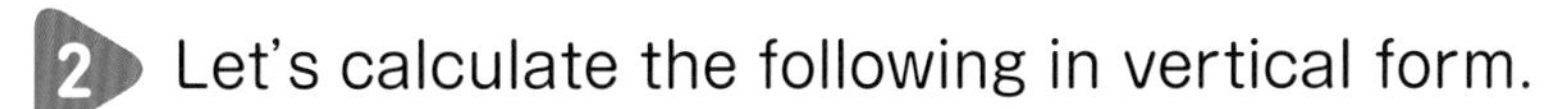

?
Can we think the same way as we have learned so far even if it is not possible to divide exactly?

There is a 2.4 m metal bar that weighs 2.72 kg. What is the weight, in kg, if the bar is 1 m long?

① Let's write a math expression.

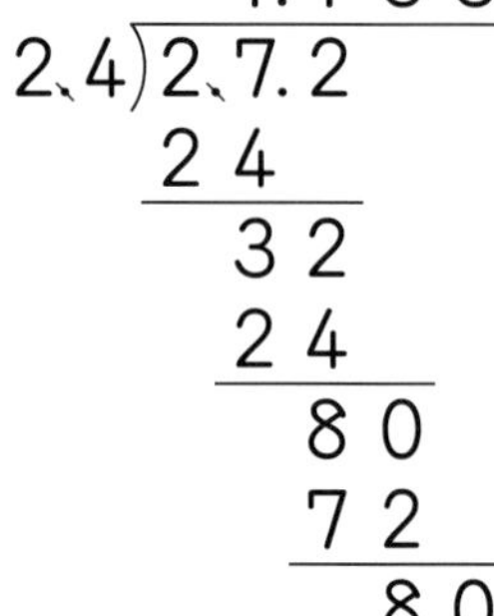

② On the right, you can see the continuation of the calculation. How can we give the answer?

Even if I continue the division, 3 will not be exactly divided.

How can we represent the answer?

\ Want to know /

? Purpose **How can we represent the quotient when the division is not exact?**

③ To find the quotient, let's round to the nearest hundredths by looking to the thousandths.

1 There is a 0.3 m wire that weighs 1.6 g. What is the weight of the wire, in g, if it is 1 m long? Let's round to the nearest tenths by looking to the thousandths.

! **The quotient can be found by rounding when the division cannot be exactly divided or when you have a bigger number of decimal places.**

2 To find the quotient, let's round to the nearest hundredths by looking to the thousandths?

① $2.8 \div 1.7$	② $6.1 \div 1.3$	③ $5 \div 2.1$
④ $61.5 \div 8.7$	⑤ $0.58 \div 2.3$	⑥ $19.2 \div 0.49$

How can we consider the remainder in a division?

6

A tape of 2.7 m long is cut into pieces that are 0.6 m long. How many of these pieces were cut and how many meters of tape remained?

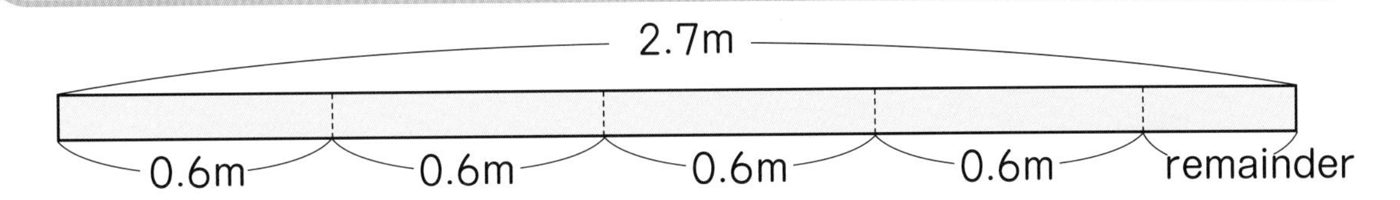

① Let's write a math expression.

② The calculation is shown below. Can you say how many meters are the remainder?

$$\begin{array}{r} 4. \\ 0.6\overline{)2.7.} \\ \underline{2\ 4} \\ 3 \end{array}$$

Can I say the remainder is 3 m?

If so, it will be larger than 0.6 m...

? **Purpose** \ Want to know / **Where should we place the decimal point of the remainder?**

③ Let's explain how much the "3 units" in the remainder represent.

! **Summary**

In divisions of decimal numbers in vertical form, the decimal point of the remainder is aligned at the same place as in the original dividend.

$$\begin{array}{r} 4. \\ 0.6\overline{)2.7.} \\ \underline{2\,4} \\ 0.3 \end{array}$$

④ Let's confirm the answer.

Dividend	=	Divisor	× Quotient	+	Remainder
2.7	=	0.6	× 4	+	

1 8 kg of rice will be placed into 1.5 kg bags. How many bags of rice will be filled? How many kilograms of rice will remain?

? We have use diagrams to think about multiplication and division, but can we rearrange them for decimals?

3 Let's think by drawing a diagram

1 A flower bed is being watered. Let's think about the area of the flower bed and the amount of water used.

I guess it depends on what you're looking for, what kind of calculation you are trying to do.

\ Want to think /

Purpose Let's draw a diagram or a table and think about what kind of calculation it will be.

① If $1\ m^2$ of the flower bed is watered with 2.4 L of water. How many liters of water are needed to water an area of $1.5\ m^2$?

Way to see and think
It is better to think of a calculation to find the total amount.

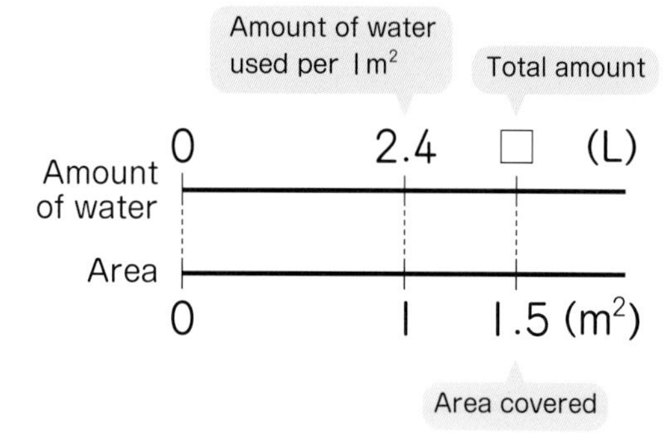

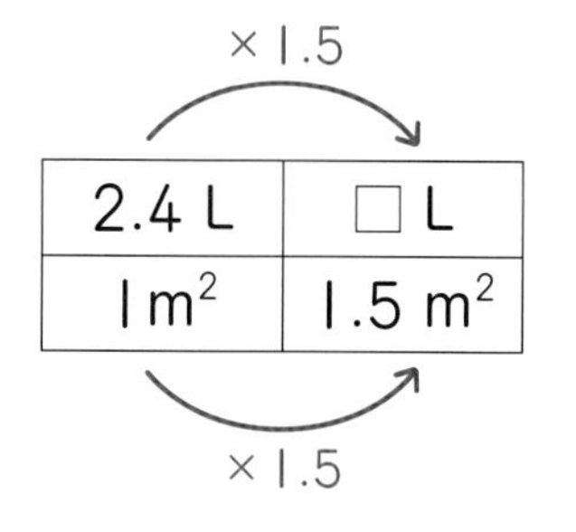

2.4 L	□ L
$1\ m^2$	$1.5\ m^2$

Math Equation: 2.4 □ 1.5 = □　　Answer: □ L

② 4 L of water were used to water an area of $2.5\ m^2$. How many liters of water are needed to water an area of $1\ m^2$?

Way to see and think
You should find the amount of water per $1\ m^2$.

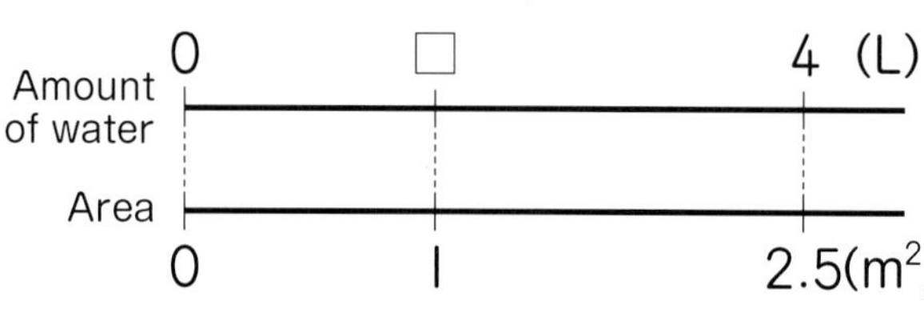

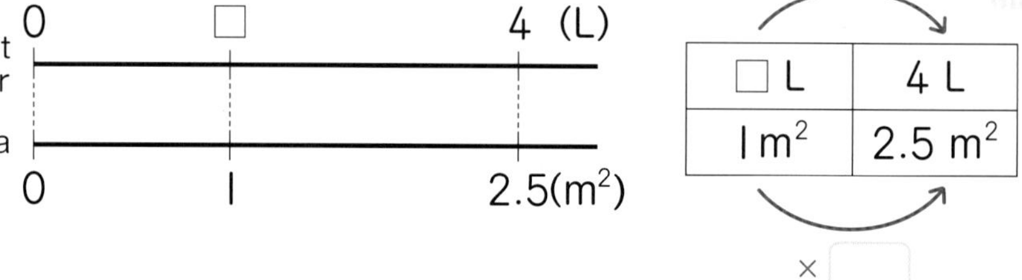

□ L	4 L
$1\ m^2$	$2.5\ m^2$

Math Equation : □ ÷ □ = □　　Answer: □ L

③ If 2.4 L of water were used to water an area of $1\,m^2$. What area of the flower bed, in m^2, can be watered with 8.4 L?

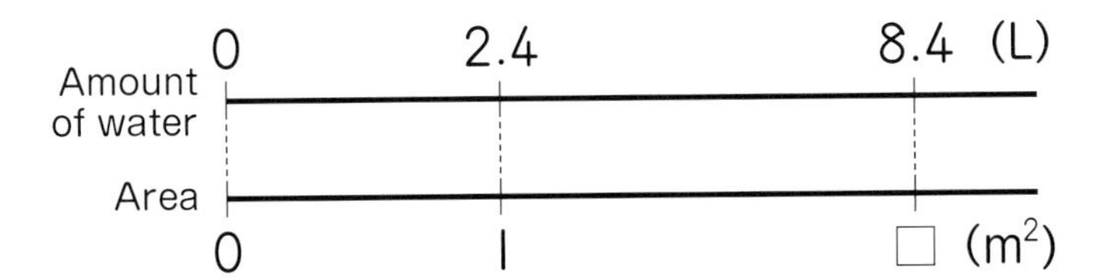

2.4 L	8.4 L
$1\,m^2$	□ m^2

Math Equation:

Answer: m^2

Summary If we know what we are looking for from this equation (Measure per unit quantity) × (how many units) = (total measurement), then we can know what kind of calculation we need to do.

Sara

1 Hikaru created the following problem.

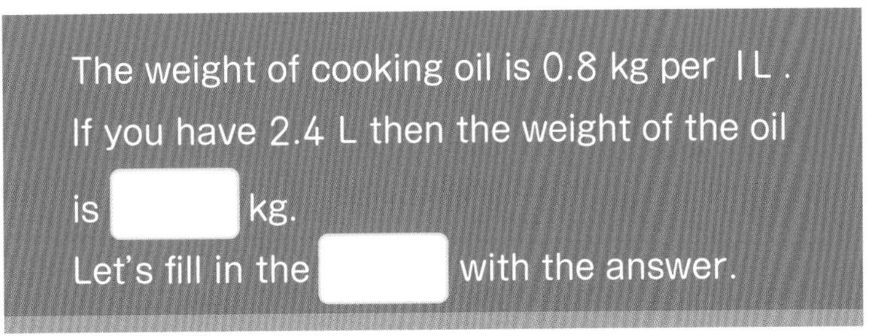

The weight of cooking oil is 0.8 kg per 1 L.
If you have 2.4 L then the weight of the oil is □ kg.
Let's fill in the □ with the answer.

① Let's find the appropriate number for □.

② Let's create a multiplication problem by changing the numbers and words.

③ Let's create a division problem by changing the numbers and words.

If it is a problem to find the total measurement, it will be a multiplication.

If we want to find the number of units, it will also be a division.

When calculating the measure per unit quantity, it is division.

I would like to make a mixed problem with decimal numbers and whole numbers.

CAN What can you do?

☐ We can calculate whole number ÷ decimal number and decimal number ÷ decimal number in vertical form. → pp.114 ~ 119

1 Let's calculate the following in vertical form. Let's continue the division until we get no remainder.

① $80 \div 3.2$　② $12 \div 0.6$　③ $39.1 \div 1.7$　④ $6.5 \div 2.6$

⑤ $29.4 \div 0.3$　⑥ $4.23 \div 1.8$　⑦ $0.99 \div 1.2$　⑧ $0.15 \div 0.08$

☐ We understand the relationship between the divisor and quotient. → p.117

2 In the following calculations, let's fill in each ☐ with the equality or inequality symbol.

① $125 \div 0.7$ ☐ 125　② $125 \div 1.3$ ☐ 125

③ $125 \div 0.89$ ☐ 125　④ $125 \div 1$ ☐ 125

☐ We can find the quotient by rounding. → p.120

3 Let's round the quotient to the nearest hundredths by looking to the thousandths.

① $0.84 \div 1.8$　② $5.18 \div 2.4$　③ $8.07 \div 0.96$

☐ We can calculate divisions with remainders. → p.121

4 Let's find the quotient as a whole number, and also give the remainder.

① $9.8 \div 0.6$　② $6.23 \div 0.23$　③ $9.72 \div 1.6$

☐ We can calculate divisions with remainders. → p.121

5 Let's find the mistake in the following calculations and write the correct answer inside the ( ).

① $4.7 \div 0.6 = 7$ remainder 5

```
       7
0.6 ) 4.7
      4 2
      ---
        5     (        )
```

② $6.2 \div 5.9 = 10$ remainder 0.3

```
       1 0
5.9 ) 6.2 0
      5 9
      -----
        3 0   (        )
```

Supplementary problems → p.162

Which "Way to See and Think Monsters" did you find in "8 Division with decimal numbers"?

Akari: When thinking about how to divide decimals, I found "Same Way."

Haruto: I found other monsters, too.

Utilize Usefulness and Efficiency of Learning

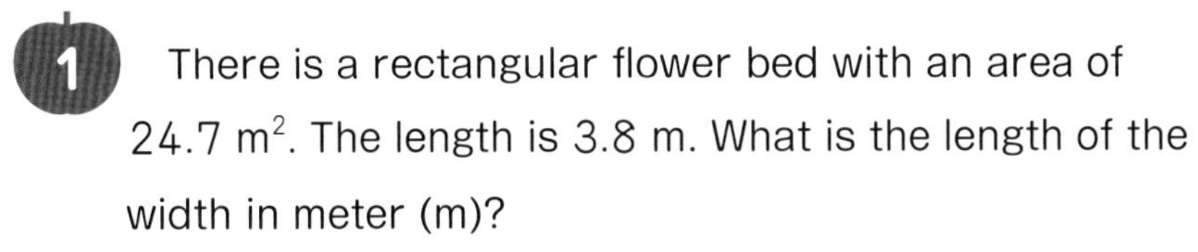

1 There is a rectangular flower bed with an area of $24.7\ m^2$. The length is 3.8 m. What is the length of the width in meter (m)?

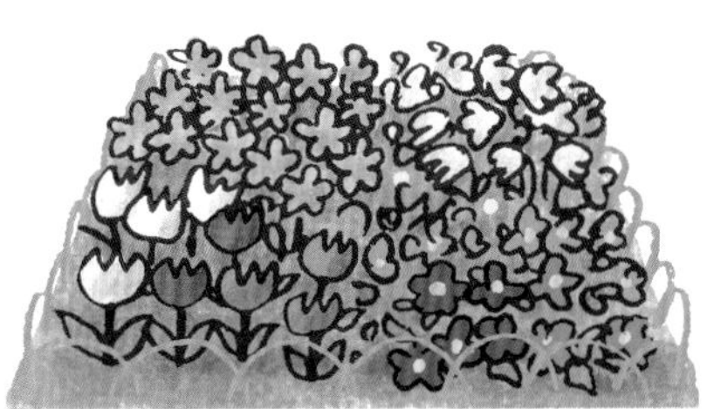

2 0.4 L of paint are used to paint a wall of $5.7\ m^2$.

① How many m^2 of the wall can you paint with 1 L?

② How many liters of paint are needed to paint if the wall is $38\ m^2$? Let's round to the nearest tenths by looking the hundredths.

3 We have 3.4 L of juice.
If we divide the juice into cups of 0.8 L each, how many cups of juice will be made and how many liters of juice will remain?

4 Let's write either the × or ÷ in the ☐ so that the math equation is correct.

① $2.8 \square 1.4 > 2.8$　　② $0.61 \square 0.4 < 0.61$

③ $7.5 \square 0.9 > 7.5$　　④ $3.58 \square 2.3 < 3.58$

5 A 4.8 m wire weighs 1.2 kg. Let's answer the following questions.

① Let's create a problem with the math expression $4.8 \div 1.2$ and find the answer.

② Let's create a problem with the math expression $1.2 \div 4.8$ and find the answer.

With the Way to See and Think Monsters...

Let's Reflect!

Let's reflect on which monster you used while learning "8 Division of Decimal Numbers."

Same Way

When dividing with decimals, we can calculate using the same way with whole numbers by converting the decimal numbers to whole numbers.

① What thoughts are needed to operate $5.76 \div 3.2$?

$$5.76 \div 3.2 = \square$$

10 times ↓　↓ 10 times

$$57.6 \div 32 = \square$$

Using the division rules, we multiplied 5.76, the dividend, and 3.2, the divisor, by 10, respectively, and converted them to decimal ÷ whole number.

$3.2\overline{)5.76}$ → $3.2\overline{)5.7.6}$ (10 times, 10 times) →

```
       1.8
3.2)5.7.6
    3 2
    ---
    2 5 6
    2 5 6
    -----
        0
```

Multiplying the dividend and the divisor by the same number does not change the quotient.

When doing the vertical form, I multiplied the dividend and divisor by 10 and then did the same calculation as learned before.

Solve the ?

Division of decimals could be converted to whole numbers and calculated in the same way as multiplication.

→

Want to Connect

Now I can calculate whole numbers and decimals, but can I also calculate fractions?

Multiplication and division of decimals are calculations that are subject to error if you do not correctly understand the relationship between the multiplicand and the product, or the dividend and the quotient. In particular, the key point is whether or not the multiplicand or the dividend is greater than 1.

① Think about the size of the quotient of □ ÷ 0.8.
The number □ is a number that is not zero. Choose one correct answer from the following Ⓐ to Ⓒ and answer its symbol.

Ⓐ The quotient of □÷0.8 will be larger than □.
Ⓑ The quotient of □÷0.8 will be smaller than □.
Ⓒ The quotient of □÷0.8 will the same as □.

The divisor is 0.8, which is smaller than 1, so...

Frequently found mistake
Consider that the quotient is smaller than the dividend.

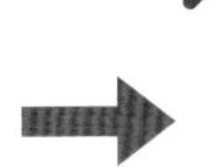

Be careful!
When the divisor is a decimal less than 1, the quotient is greater than the dividend. It will be easier to understand if you actually put a number in the □.
(Example) If you apply 8 into □, 8÷0.8=10

② In the following four math expressions, ○ represents the same number that is not zero. Choose all the options from the following Ⓓ to Ⓖ whose answer is larger than the number of ○.

Ⓓ ○× 1.2 Ⓔ ○× 0.7 Ⓕ ○÷ 1.3 Ⓖ ○÷0.8

Frequently found mistake
Consider that the product is greater than the multiplicand and the quotient is less than the dividend.

Be careful!
In multiplication, the product is greater than the multiplicand when the multiplier is a decimal greater than 1.
Conversely, in division, when the divisor is a decimal less than 1, the quotient is greater than the dividend.

If you are not sure, try to think by applying an actual number.
For example, if you put 10 in ○,
Ⓓ 10 × 1.2 = 12 → Larger than 10
Ⓔ 10 × 0.7 = 7 → Smaller than 10
Ⓕ 10 ÷ 1.3 = 7.69⋯ → Smaller than 10
Ⓖ 10 ÷ 0.8 = 12.5 → Larger than 10

How many times: Decimal multiples

Let's compare lengths.

1 There are four types of tapes, ⓐ to ⓓ, as shown on the right. Answer the following questions about the lengths of these tapes.

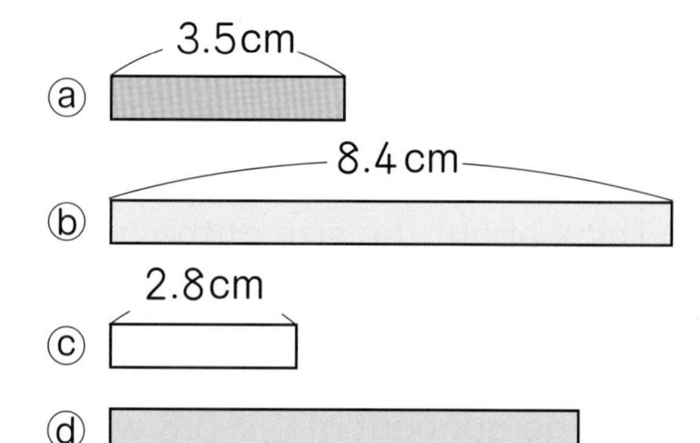

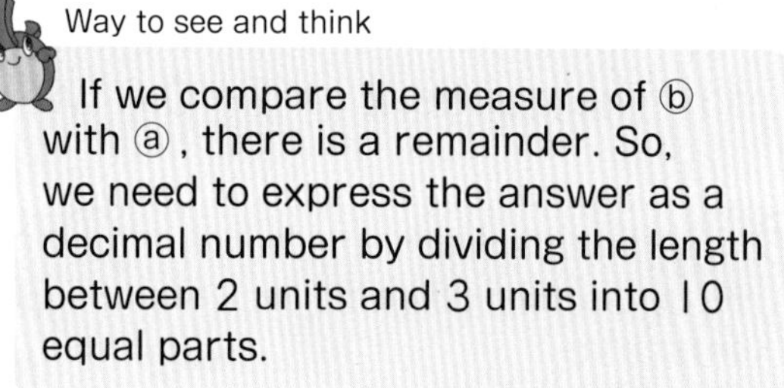

If we compare the measure of ⓑ with ⓐ, there is a remainder. So, we need to express the answer as a decimal number by dividing the length between 2 units and 3 units into 10 equal parts.

① How many times the length of ⓑ is the length of ⓐ?

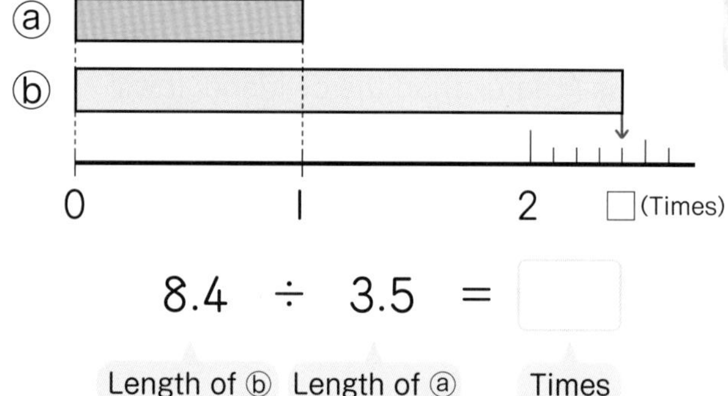

$8.4 \div 3.5 = \square$

Length of ⓑ　Length of ⓐ　Times

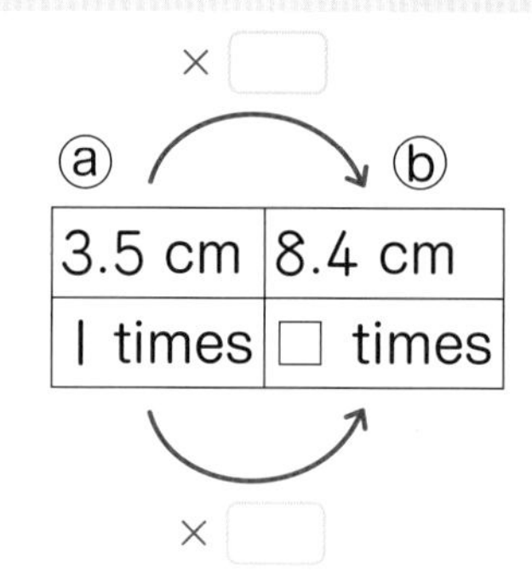

3.5 cm	8.4 cm
1 times	□ times

\ Want to explore /

Purpose Can you find the relationship between the different multiples?

② How many times the length of ⓒ is the length of ⓐ?

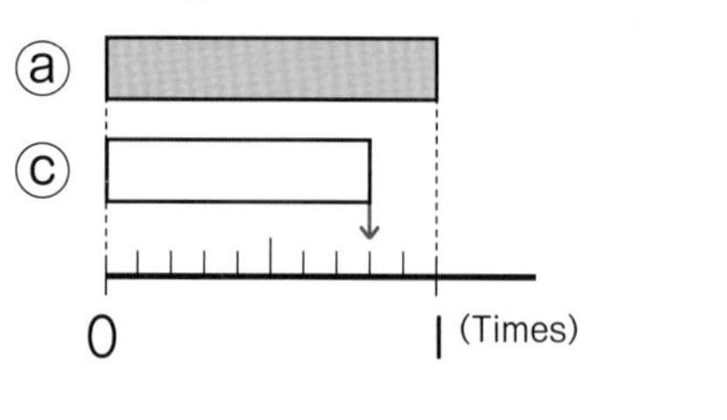

$2.8 \div 3.5 = \square$

Length of ⓒ　Length of ⓑ　Times

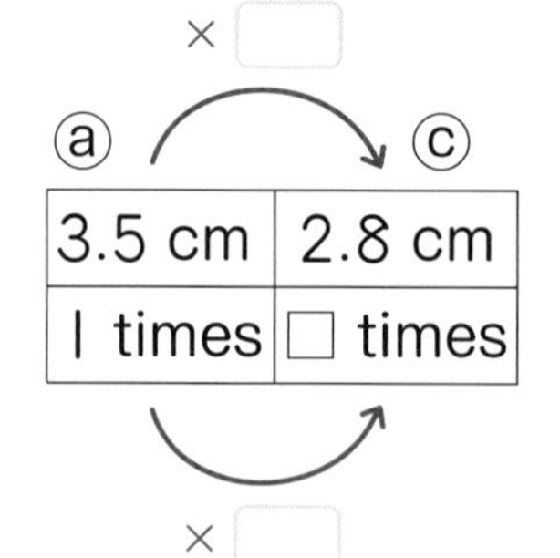

3.5 cm	2.8 cm
1 times	□ times

③ The length of ⓓ is 2.5 times the length of ⓒ. What is the length of ⓓ in cm?

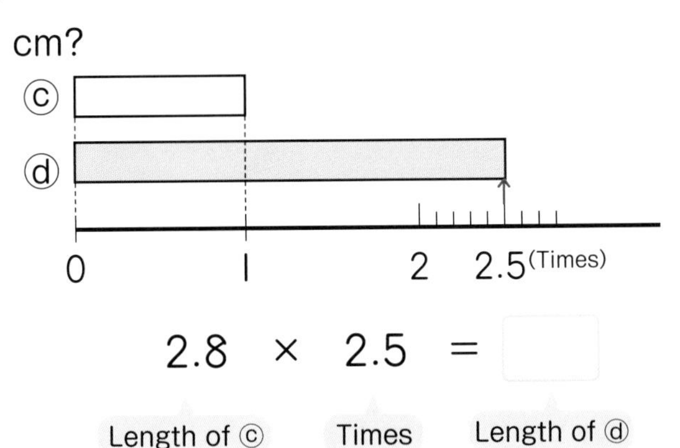

ⓒ	ⓓ
2.8 cm	□ cm
1 times	2.5 times

×2.5 (ⓒ → ⓓ)

×2.5 (1 times → 2.5 times)

$2.8 \times 2.5 = \square$

Length of ⓒ　Times　Length of ⓓ

Summary When the original size is changed, the relationship of times also changes. We have to pay attention to what the original size is.

④ How many times the length of ⓒ is the length of ⓓ?

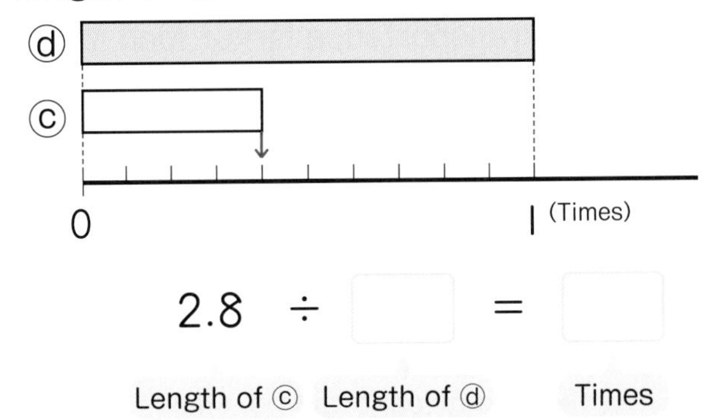

$2.8 \div \square = \square$

Length of ⓒ　Length of ⓓ　Times

ⓓ	ⓒ
7 cm	2.8 cm
1 times	□ times

× □ (ⓓ → ⓒ)

× □ (1 times → □ times)

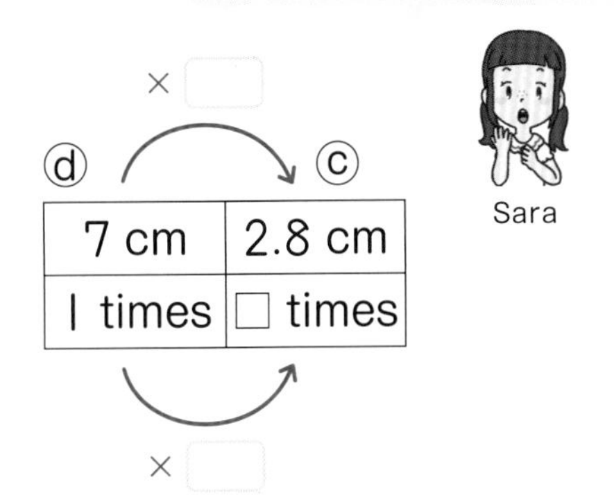

1 The length of red tape is 120 cm. The length of the red tape is 0.6 times the length of the white tape. Which diagram represents correctly the relationship between the length of the red and the white tapes? Let's choose among ⓔ~ⓗ.

What is the original size?

Yu

ⓔ

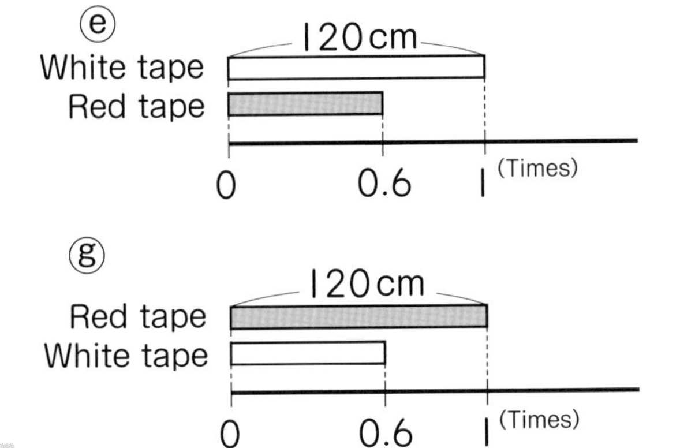

ⓕ

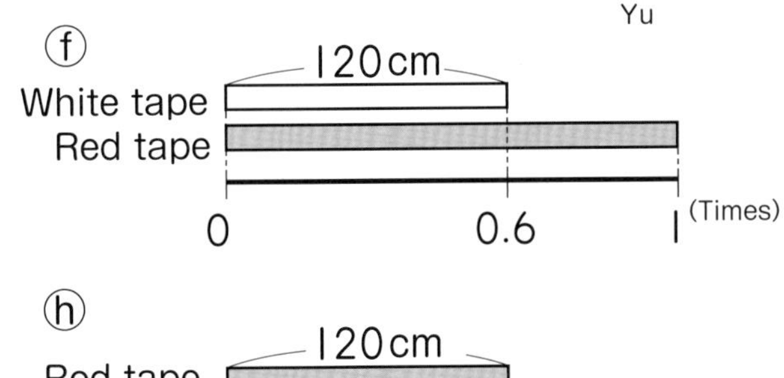

ⓗ

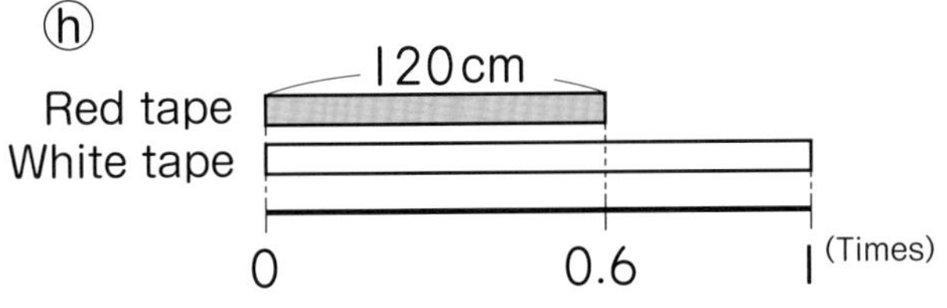

Utilizing Math for SDGs

Let's think about the environment from the perspective of food.

Have you ever heard of the term "food mileage"? It is calculated by multiplying the weight of food transported (in tons) by the distance transported (in kilometers), and the unit is "t-km (ton-kilometer)."

It is used as a rough guide when considering the environmental impact of transportation.

For example, if the same amount of food is transported, a larger food mileage indicates that the food is produced and transported from farther away, while a smaller food mileage indicates that the food is produced and transported from a closer place.

Since Japan imports a lot of food, the food mileage per year is considerably larger than in other countries, and is a major burden on the environment. In order to reduce this burden, the concept of "Chisan Chisho (local production for local consumption)," is a very important concept. The merits of Chisan Chisho are as follows; can obtain new agricultural products and marine products; can maintain our traditional food culture by using local ingredients; can save the time and labor of transportation. Some school lunch programs are trying to increase the use of local products.

Let's think about the relationship between the food around us and the environment.

Within Japan, the transportation time is short.
→ Food mileage will be small.

When imported from abroad, it takes a long time to transport.
→ Food mileage is high.

① If a bakery uses 300 kg of flour per day, how much difference is there in food mileage between flour produced in Hokkaido and flour produced in the U.S.A.?

Place of production	Hokkaido	The United States
Distance from the place of production from the bakery (Tokyo)	1118 km	19885 km

② Find out where the ingredients used in your school lunch come from. Let's also look into locally produced ingredients and see if we can make use of them in our school lunches.

On the school lunch menu, there is a mark that represents "local production for local consumption."

Which materials are locally made?

Think back on what you felt through this activity, and put a circle.

Let's reflect on yourself!

	Strongly agree	Agree	Do not agree
① I could find out information about "Chisan Chisho."			
② I could consider about food mileage.			
③ I could utilize the knowledge of math to find out the food mileage.			

	Strongly agree
④ I am proud of myself because I did my best.	

Let's praise yourself with some positive words for trying hard to learn!

Find the ?

What is the measure of the angles of the triangle rulers?

We learned each of the sizes of angles in a triangle ruler.
The size of each angle is 45°, 45,° and 90° .
The sizes for these angles are 30°, 60,° and 90° .
Both have right angles.
1
Ah? If you add all the sizes of the angles, on either of the triangles, it becomes 180°.
45°
45°
90°
60°
30°
90°
2
I wonder if it would be the same with the other triangles...
3

\ Want to explore /

Will the internal angle sum of a triangles always be the same?

9 Angles of Figures

Let's explore about the angles of triangles and quadrilaterals.

1 Internal angle sum of a triangle

1 For the right triangle shown on the right, angle A becomes smaller from 60°, 50°, 40°, ... and so on. So, vertex B gets closer to vertex C. Let's examine the measure of the angles when this happens.

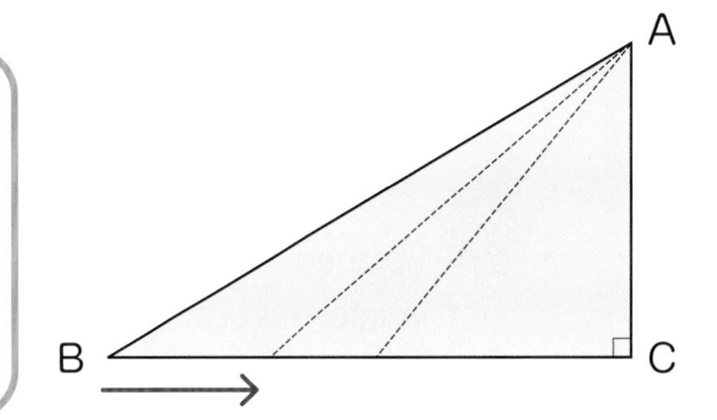

① How does the size of angle B change when the measure of angle A decreases by 10°? Let's measure angle B with a protractor and organize the information in the table below.

② Let's find the sum of angle A and angle B.

Angle A (°)	60	50	40			
Angle B (°)						
Sum (°)						

Akari

As angle A gets smaller, angle B gets bigger.

Even if the measure of angle A changes, there are things that do not change.

Yu

③ Let's discuss what happens to the sum of the three angles of a triangle?

Sara

It looks like the sum of the 3 angles of a right triangle is 180°.

I wonder if the same can be said for other triangles.

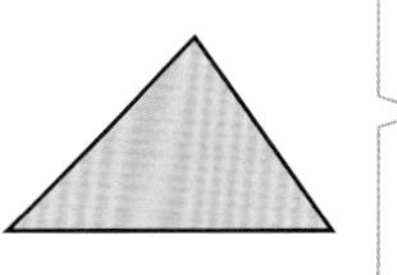

Haruto

④ Let's draw triangles by ourselves, and investigate the sum of the three angles of the triangles.

Let's think about how to measure without using a protractor.

1 Let's explain the ideas of the following children.

The measure of a straight line angles is 180°.

Akari's idea

I cut the 3 angles and rearranged each vertex into one.

Since all 3 rearranged angles became a straight line, their sum is ☐°.

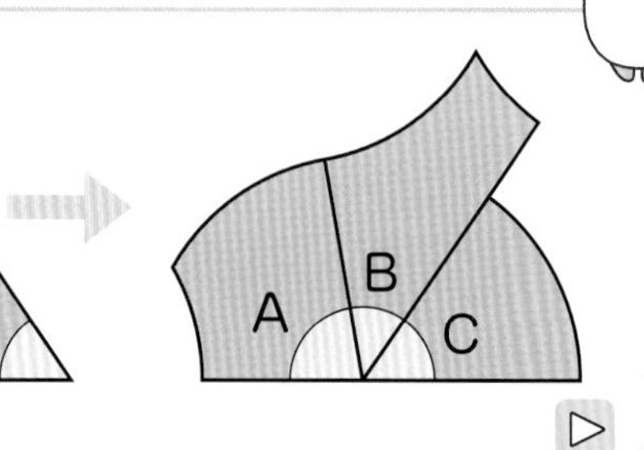

Yu's idea

I placed congruent triangles together, without any gap, to form a continuous pattern.

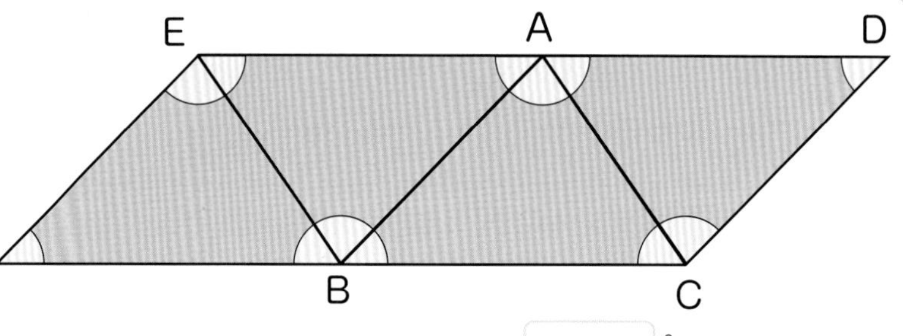

Since the 3 angles that meet at A or B became a straight line, their sum is ☐°.

Sara's idea

At which points on sides AB and AC is it folded?

I folded the triangle and attached together the 3 angles.

Since all 3 angles became a straight line, their sum is ☐°.

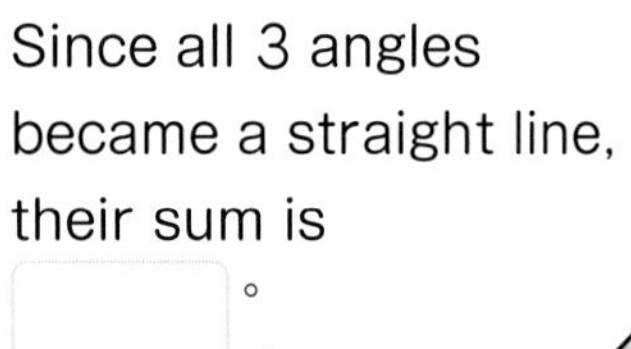

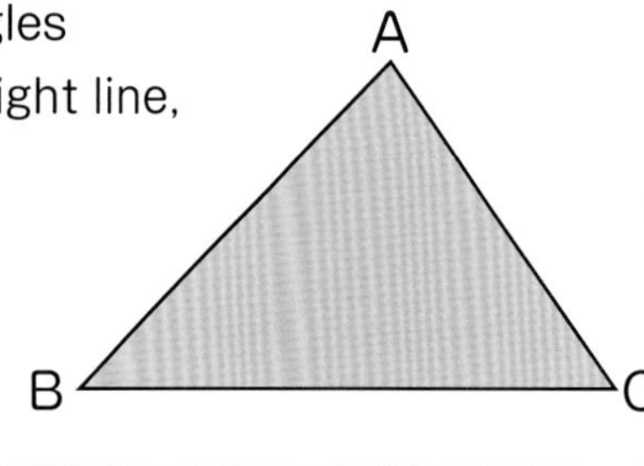

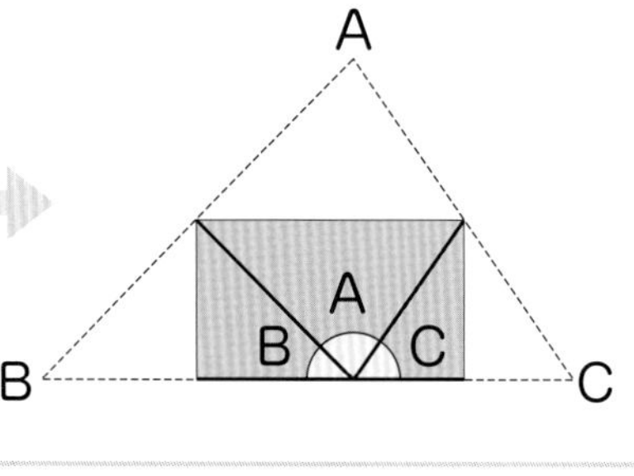

Summary

The sum of the three angles of any triangle is 180°.

Way to see and think

The rule can be confirmed in different ways.

? Using the fact that the sum of the three angles of a triangle is 180°, can you find the angles of various triangles?

2 Let's find the measure of the following angles Ⓐ~Ⓕ by doing calculations.

\ Want to try /

Purpose Think about it by using the fact that the sum of the three angles of a triangle is 180°.

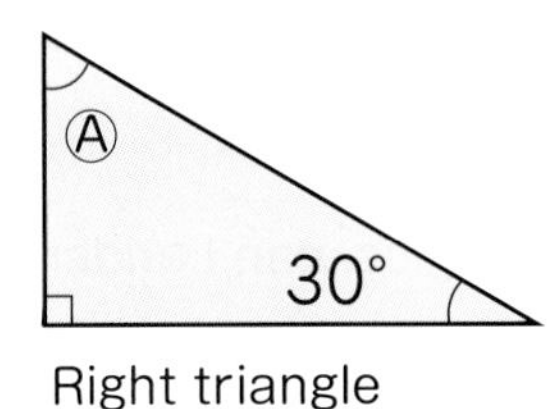

Right triangle

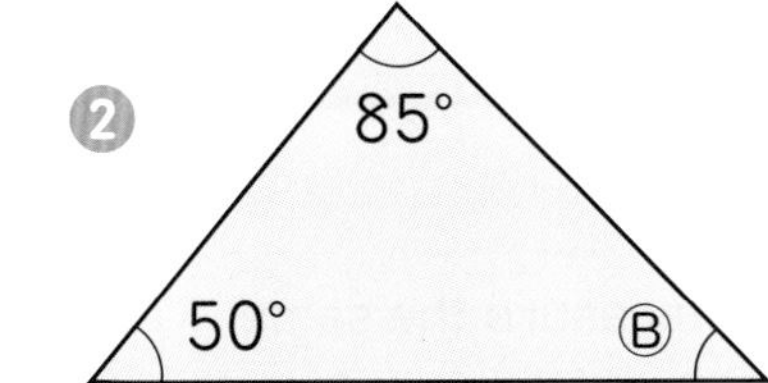

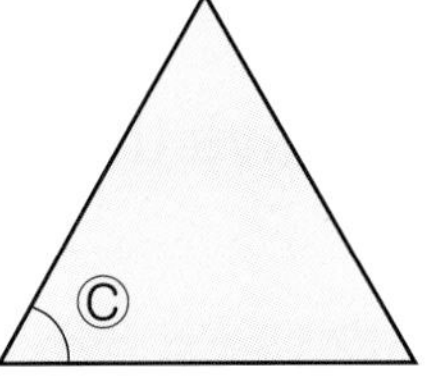

Equilateral triangle

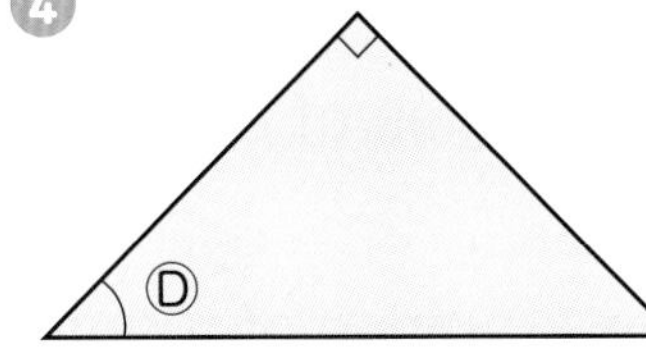

Isosceles right triangle

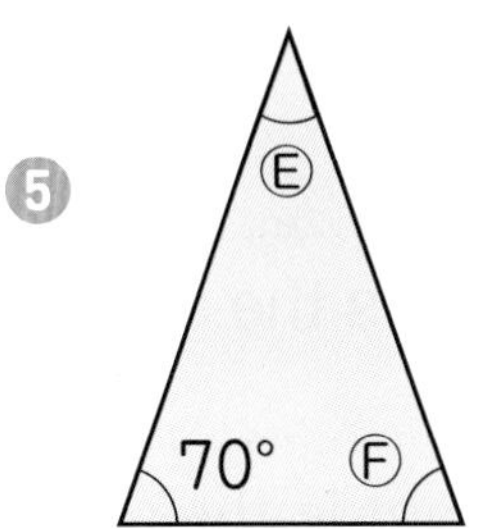

Isosceles triangle

1 Let's think about the triangle shown on the right.

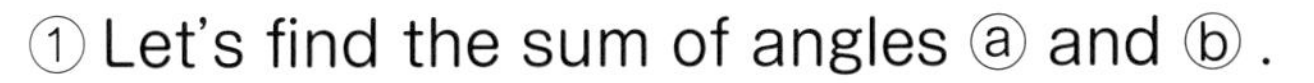

① Let's find the sum of angles ⓐ and ⓑ.

② Let's find the measure of angle ⓒ.

③ What can you conclude about the sum of angles ⓐ, ⓑ, and ⓒ?

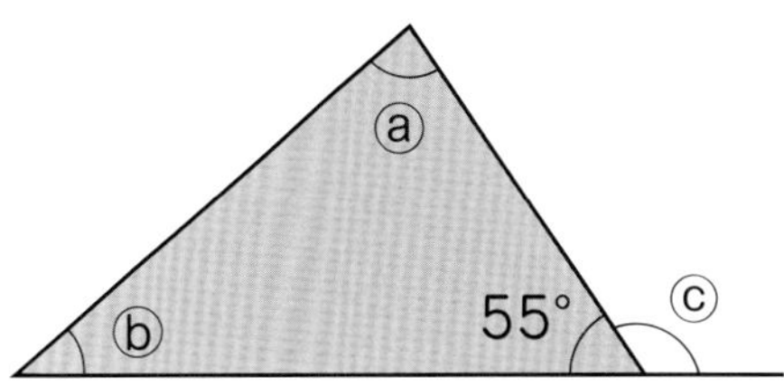

2 Let's find the measure of angles ⓐ~ⓒ by doing calculations.

①

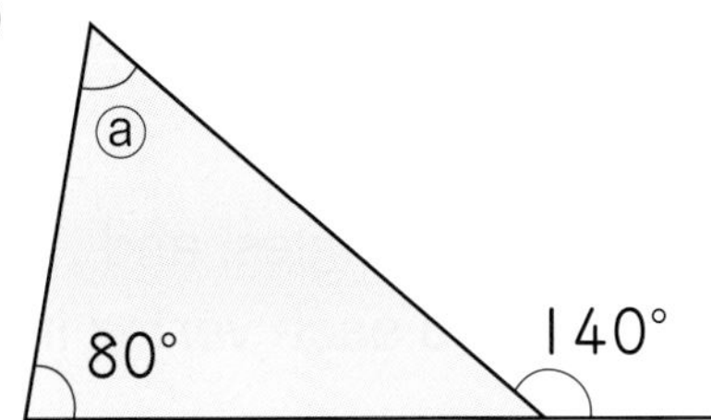

②

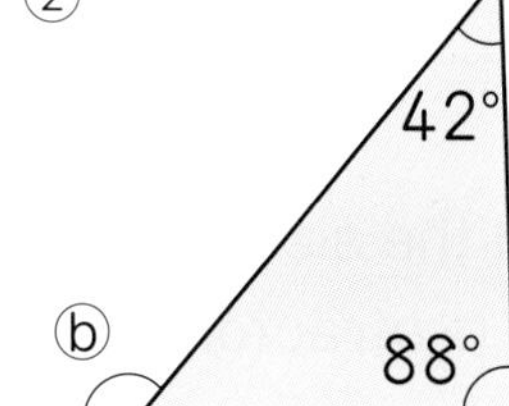

③

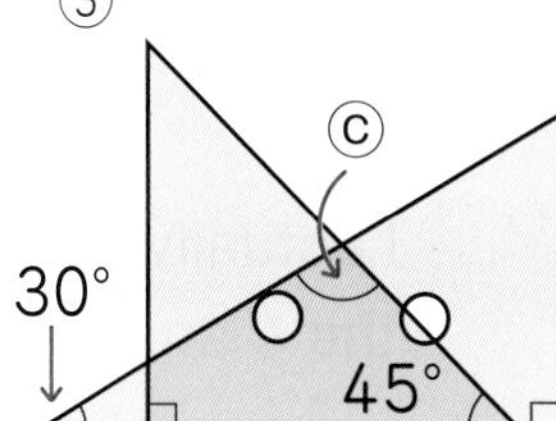

? Is there any rule for the internal angle sum of a quadrilateral?

2 Internal angle sum of a quadrilateral

1 Let's explore about the sum of the four angles of a quadrilateral.

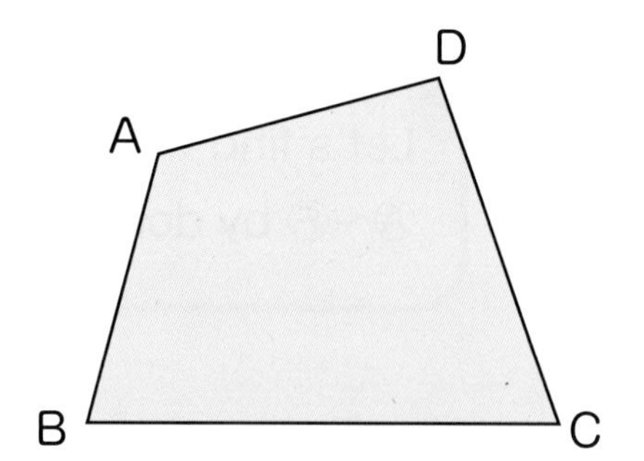

Should I try to measure the same way as I did with the triangle?

As in triangles, can I understand quadrilaterals if I connect the angles?

\ Want to explore /

Is the sum of the 4 angles of a quadrilateral always the same?

① What is the sum of the 4 angles of a quadrilateral? Let's explore in different ways.

Way to see and think

Can it be examined in the same way as what we did with the angle sum of triangles?

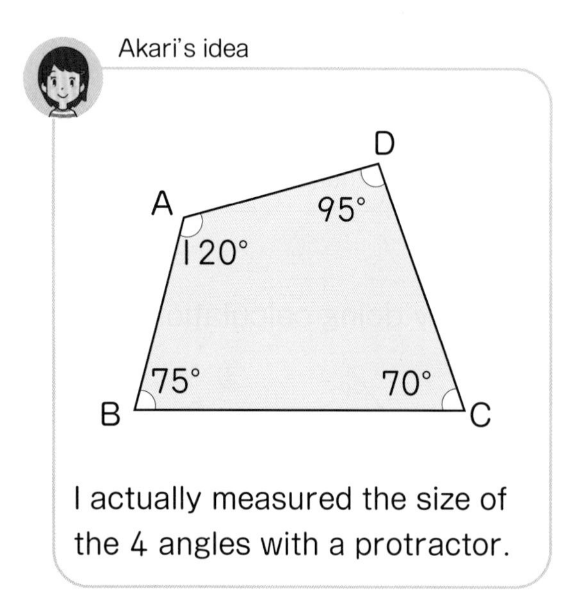

Akari's idea
A
D
95°
120°
75°
70°
B
C
I actually measured the size of the 4 angles with a protractor.

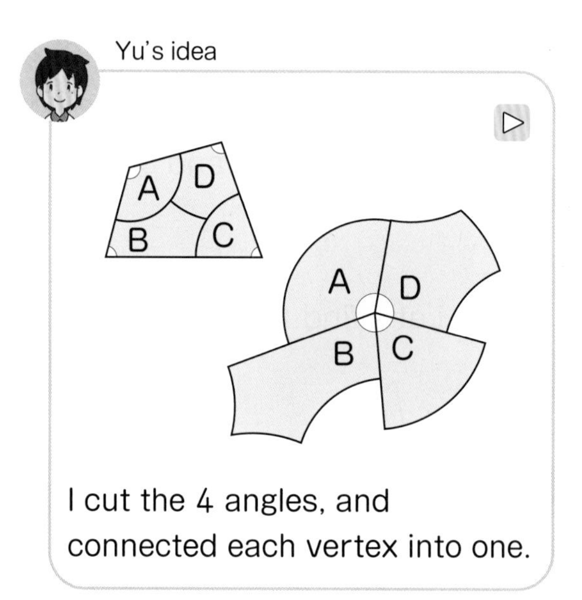

Yu's idea
A
D
B
C
A
D
B
C
I cut the 4 angles, and connected each vertex into one.

② Let's discuss what other methods are possible.

③ Let's explain the ideas of the following children.

Sara's idea

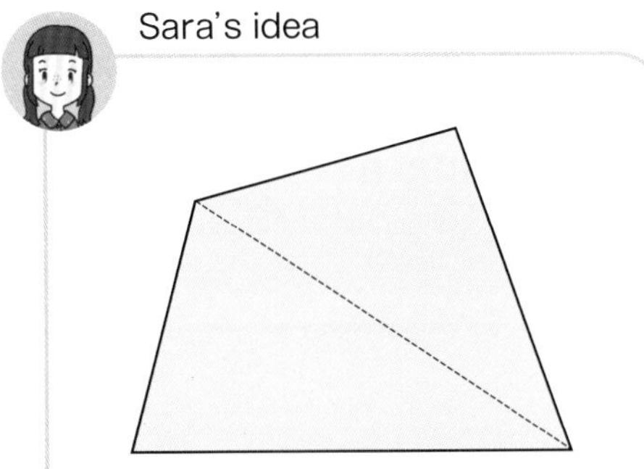

Divide the quadrilateral with 1 diagonal. Since 2 triangles are formed,

$\square^\circ \times 2 = \square^\circ$

Way to see and think

Thinking based on the angle sum of the 3 angles of a triangle.

Haruto's idea

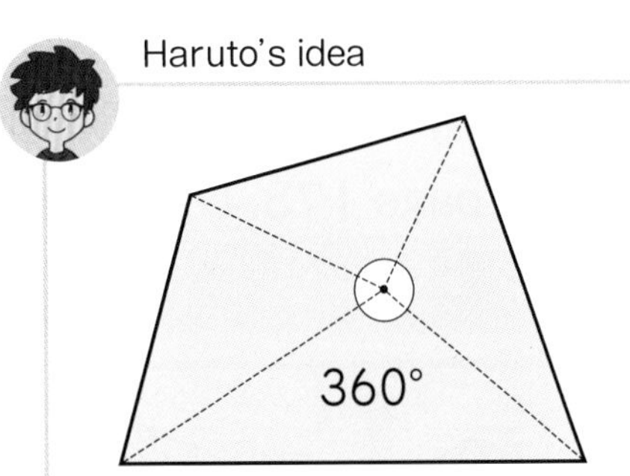

Take a point inside the quadrilateral, and divide it into 4 triangles. Since 4 triangles are formed,

$\square^\circ \times 4 = \square^\circ$

Subtract the angle that is formed at that point $\square^\circ$

So, $\square$.

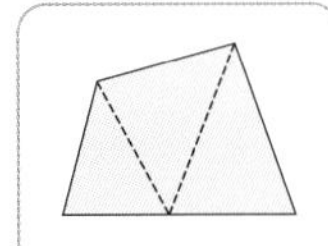

I can divide it like this.

Akari

Summary

The sum of the 4 angles of any quadrilateral is 360°.

1 Using the tessellation of congruent quadrilaterals, let's confirm that the sum of the 4 angles of a quadrilateral is 360°.

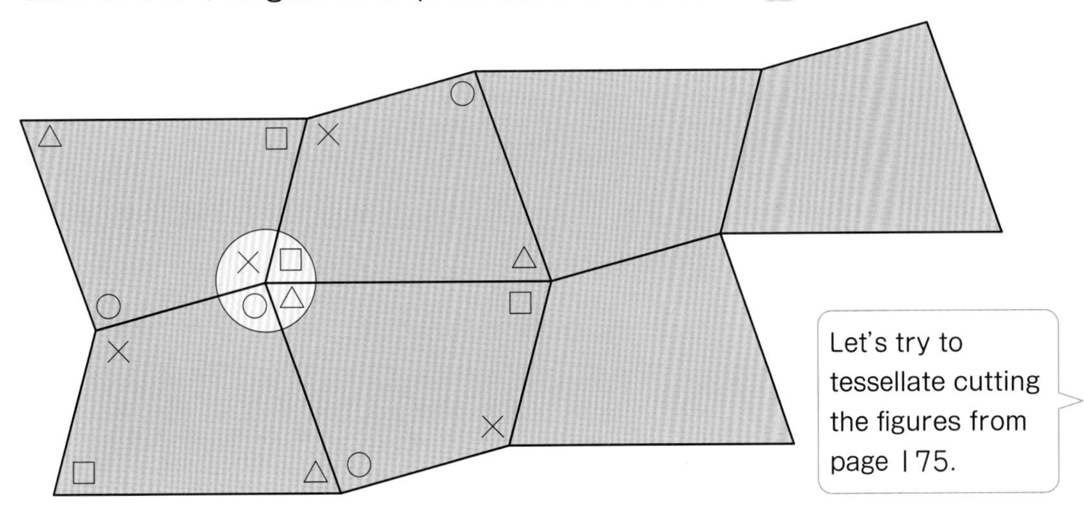

Let's try to tessellate cutting the figures from page 175.

? Can we make sure that the sum of the 4 angles of any quadrilateral is 360° ?

2 I would like to verify that the sum of the 4 angles of the figure shown on the right is also 360°. Cut the figures from page 175. Let's explore by creating a tessellation.

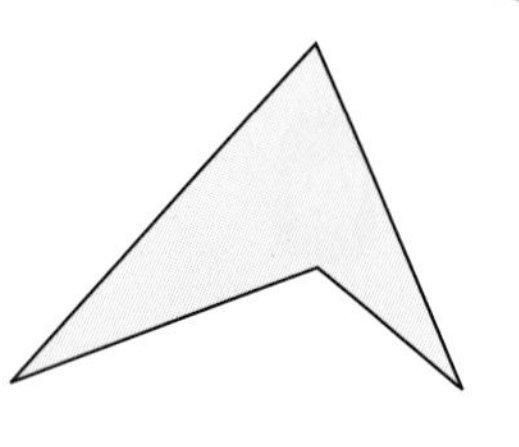

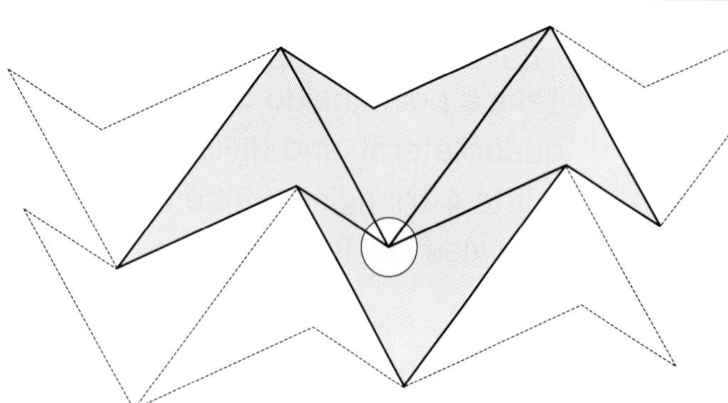

\ Want to think /

Purpose What kind of angle is made in that point?

Haruto

1 The verification to see if the sum of the 4 angles of the figure **2** is 360° was done as follows. Let's explain the ideas of the following children.

Akari's idea

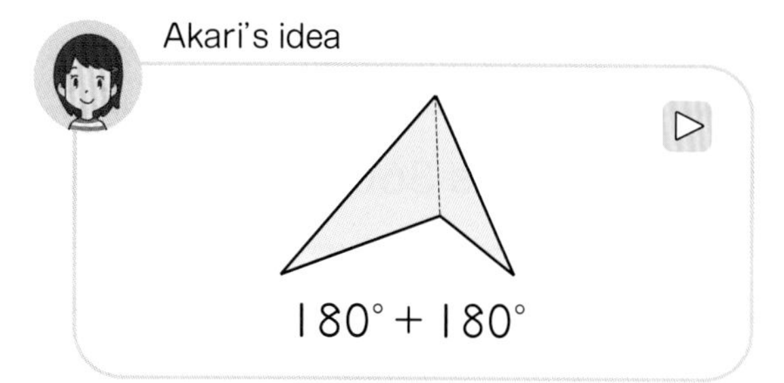

Yu's idea

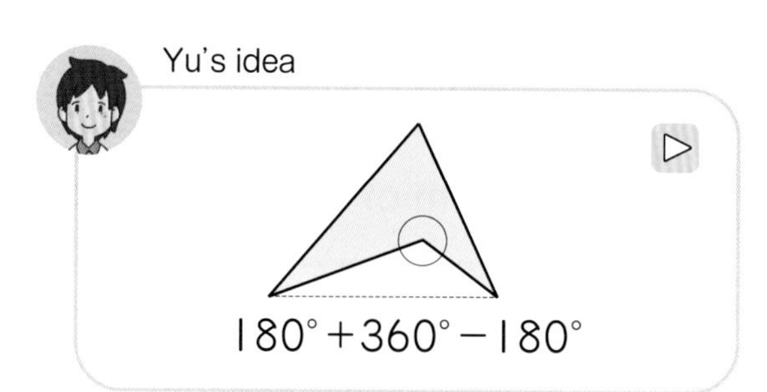

Summary If we close the quadrilateral, all 4 angles come together at one point, so we get 360°.

Sara

2 Let's find the measure of the following angles ⓐ~ⓒ by doing calculations.

①

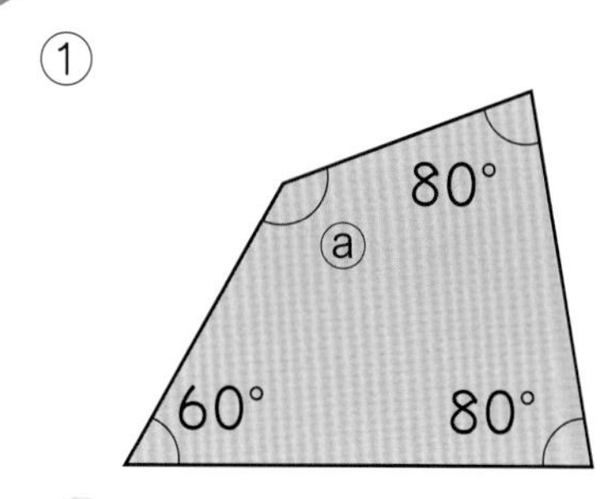

②

100°
ⓑ

③

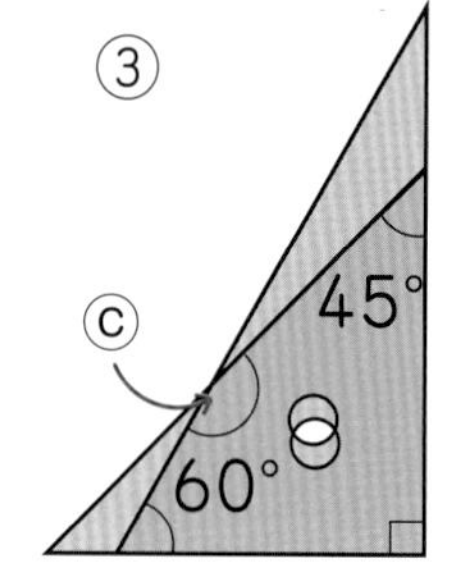

? Is there any rule for the internal angle sum of other shapes?

A figure enclosed by 5 straight lines is called a pentagon. Let's think about how to find the sum of the 5 angles of a pentagon.

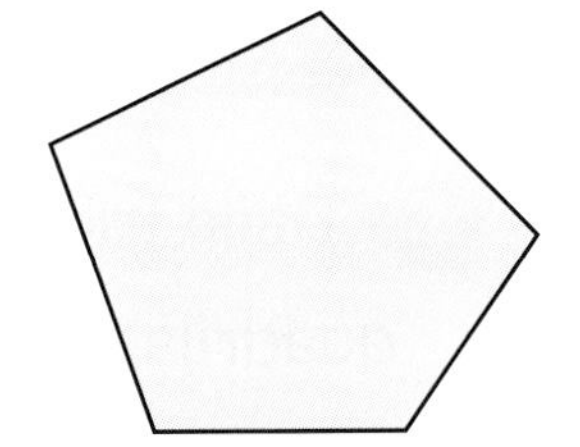

Haruto: Can I think about it as the tessellations we did before?

Akari: What if I divide into triangles, as we did with the quadrilaterals?

\ Want to explore /

? **Purpose** What is the angle sum of the 5 angles of a pentagon?

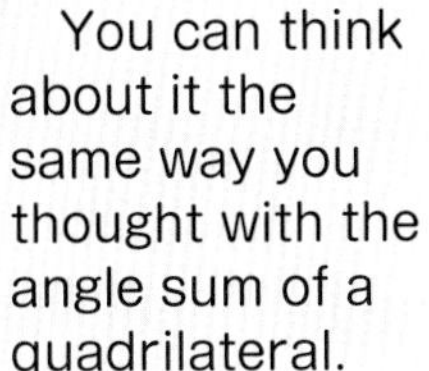

Way to see and think

You can think about it the same way you thought with the angle sum of a quadrilateral.

① Let's explain the ideas of the following children.

Yu's idea

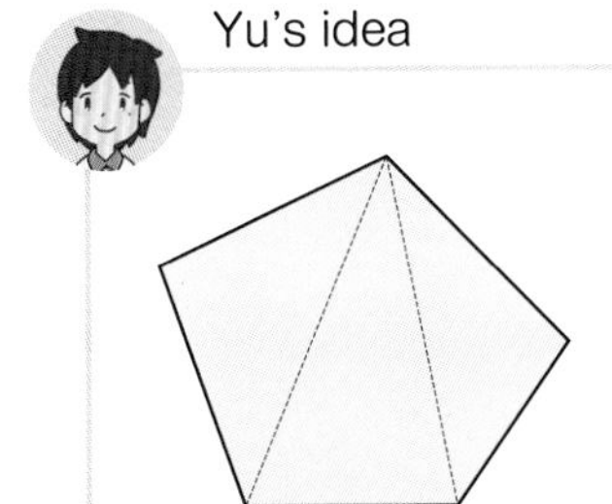

If you make straight lines from one vertex to another that is not adjacent, it can be divided into ☐ triangles.

Therefore, $180^\circ \times$ ☐ $=$ ☐ $^\circ$

Sara's idea

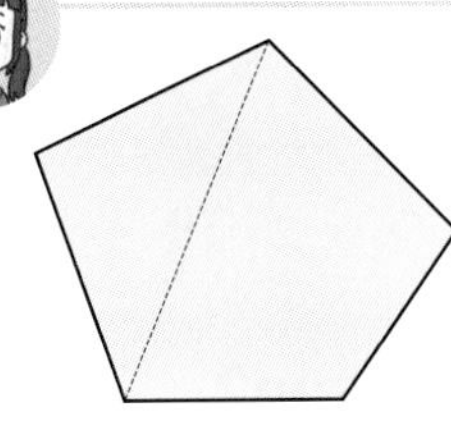

If you make only one straight line from one vertex to another that is not adjacent, then it will be divided into a triangle and a quadrilateral.

Therefore, $180^\circ +$ ☐ $=$ ☐ $^\circ$

Akari's idea

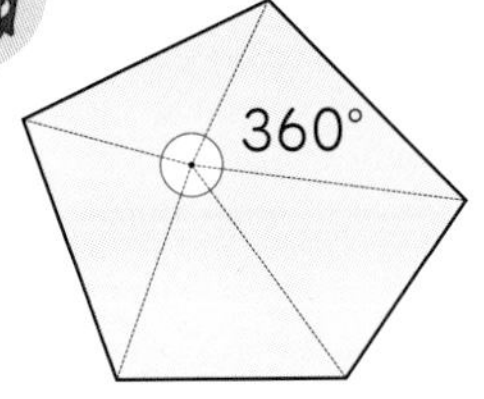

Take a point inside the pentagon and divide it into 5 triangles. Since 5 triangles are created,

☐ $^\circ \times 5 =$ ☐ $^\circ$

Then, subtract the angle that is formed at that point ☐ $^\circ$.

Therefore, ☐ $^\circ$.

Summary

The sum of the 5 angles of any pentagon is 540°.

1 Daiki tried to tessellate the pentagon the same way as we did with the quadrilaterals, but he could not do it. Let's explain the reason why it was not possible.

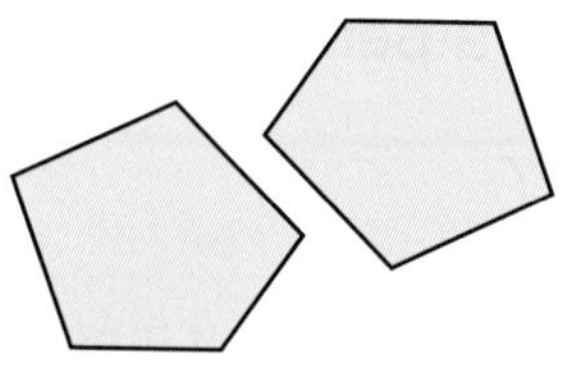

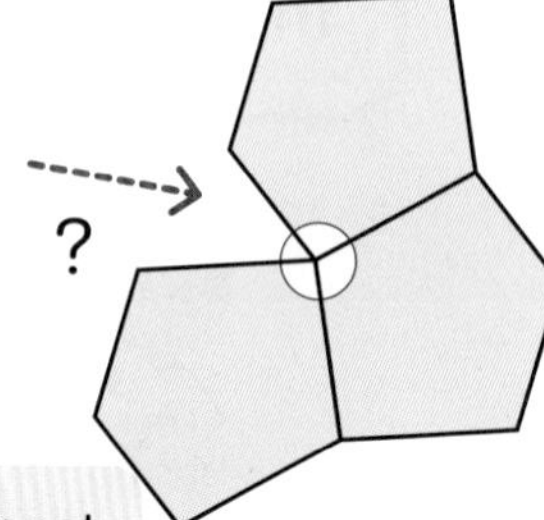

If you want to tessellate, the sum of the angles connected must be ....

Way to see and think

What happens if you connect 5 angles at one vertex...

? Is there a rule for the size of angles in shapes enclosed by more straight line?

Angle sum of polygons →

2 **A figure enclosed by 6 straight lines is called a hexagon. Let's think about how to find the angle sum of the 6 angles of a hexagon.**

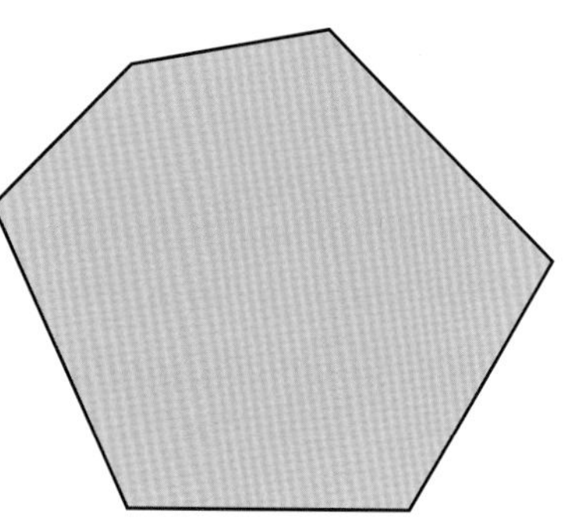

It seems maybe the same idea as before can be use.

Can I find any pattern to create a rule?

Want to explore

? **Purpose** What is the angle sum of the 6 angles of a hexagon?

① Let's explain how to find the angle sum of the 6 angles of a hexagon by using the figure shown above.

Summary

The sum of the 6 angles of any hexagon is 720°.

A figure that is enclosed only by straight lines, such as triangles, quadrilaterals, pentagons, hexagons, etc., is called a **polygon.** In a polygon, each straight line that connects any two vertices that are not adjacent is called a **diagonal**.

1 Let's summarize the angle sum of different polygons using the number of triangles that can be created by drawing diagonals from only one vertex.

	Triangle	Quadrilateral	Pentagon	Hexagon	Heptagon	Octagon	Nonagon
Number of triangles	(1)	2	3	4			
Angle sum	180°	360°	540°	720°			

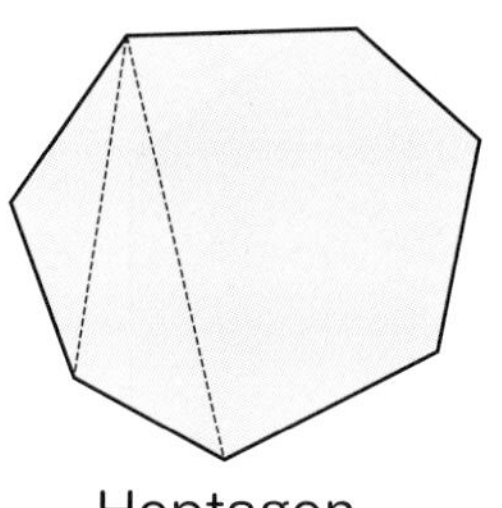

Heptagon

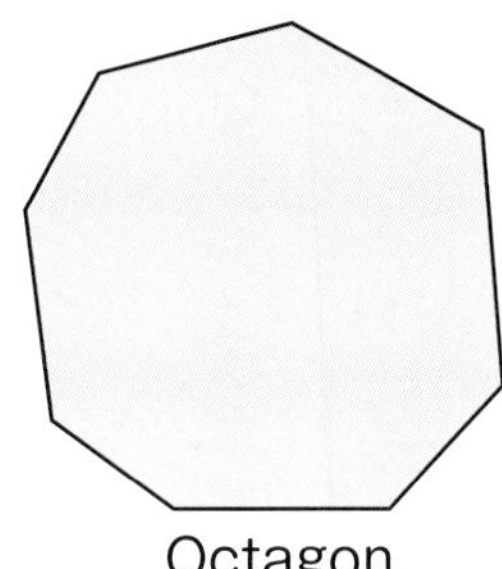
Octagon

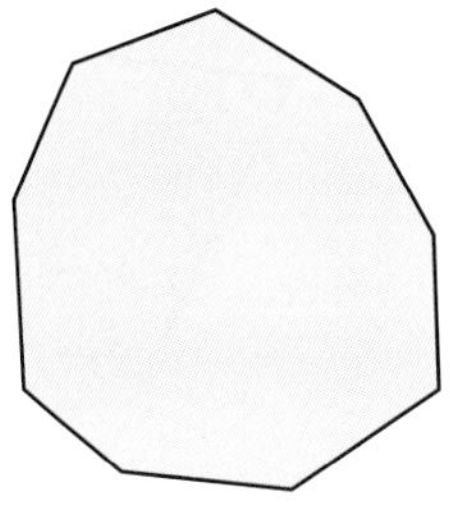
Nonagon

180° × ☐ = ☐ °　180° × ☐ = ☐ °　180° × ☐ = ☐ °

Way to see and think

How does the angle sum changes as the number of triangles increases?

2 The following math expressions represent how to find the angle sum of a pentagon. Match the figure from ⓐ~ⓓ with the corresponding math expression from ①~④.

① 180° + 360°　② 180° × 3

③ 180° × 4 − 180°　④ 180° × 5 − 360°

ⓐ 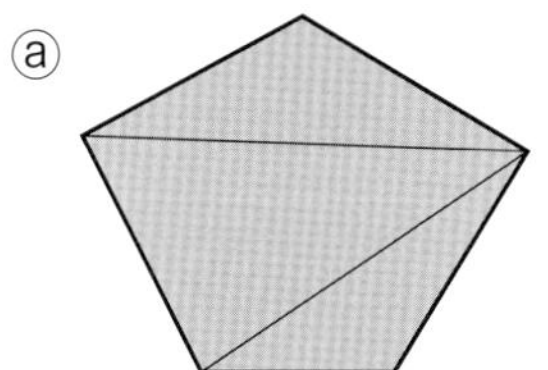

ⓑ 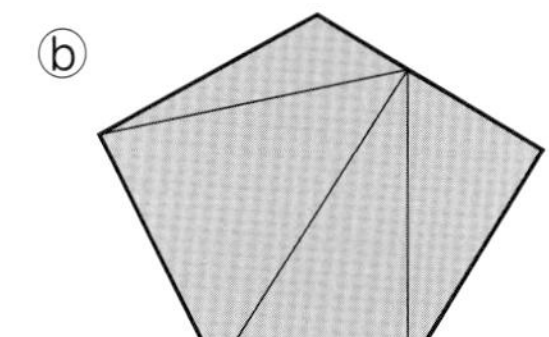

ⓒ 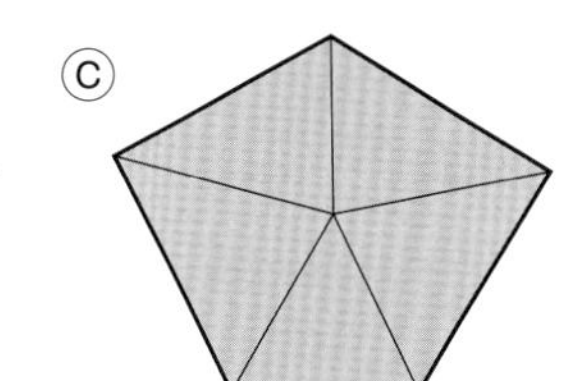

ⓓ 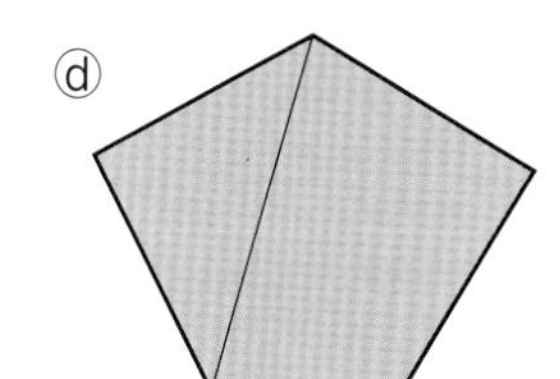

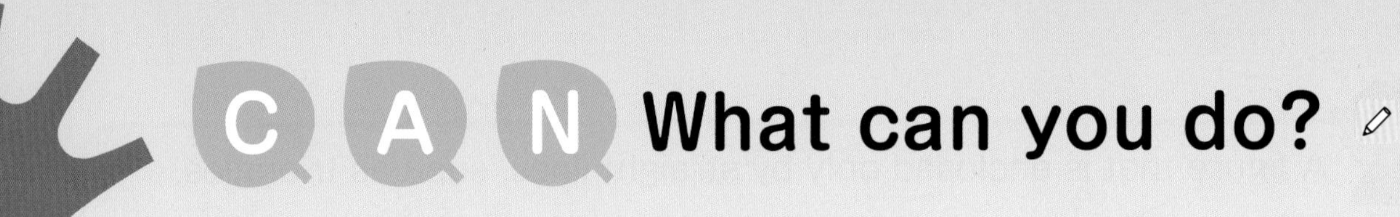

What can you do?

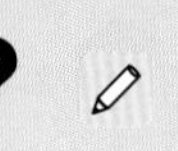

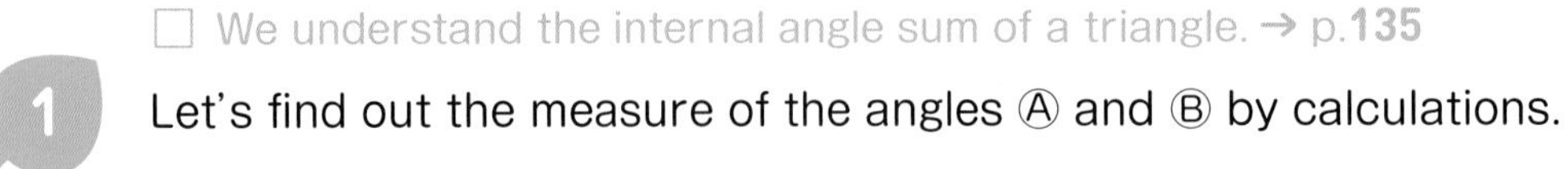

☐ We understand the internal angle sum of a triangle. → p.135

1 Let's find out the measure of the angles Ⓐ and Ⓑ by calculations.

①

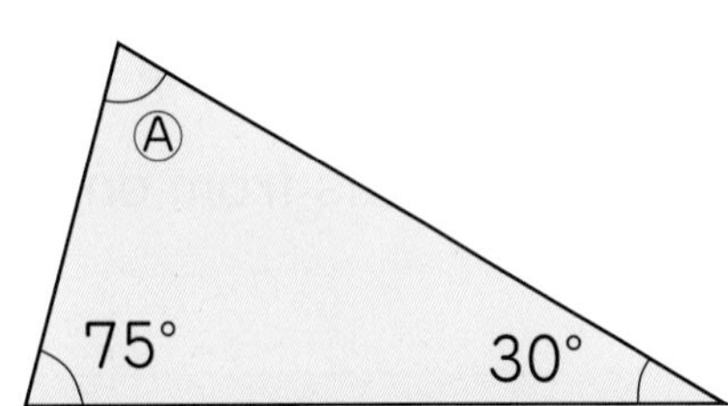

②

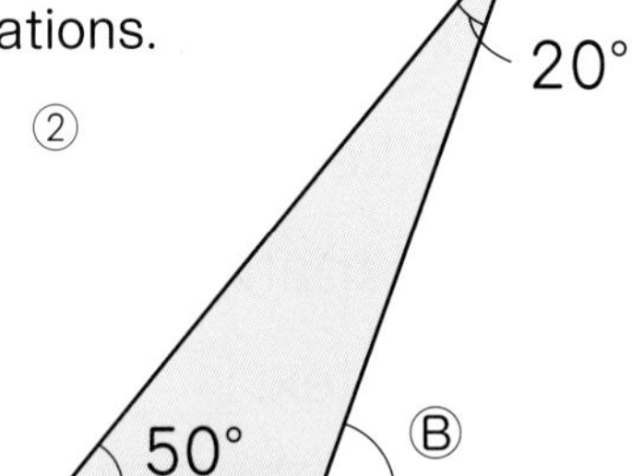

☐ We understand the internal angle sum of a quadrilateral. → pp.136 ~ 138

2 Let's find out the measure of angles Ⓐ and Ⓑ by calculations.

①

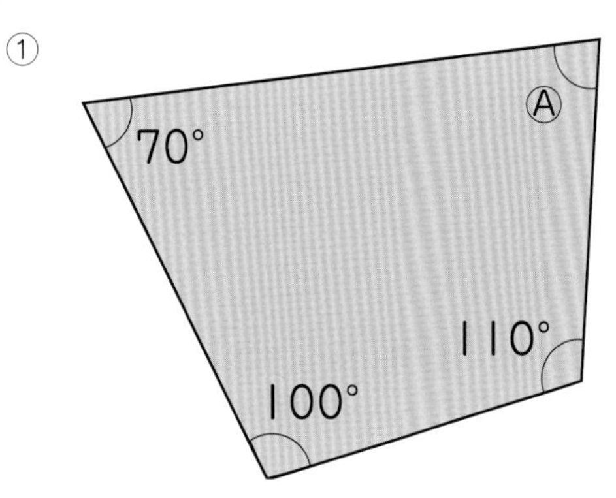

②

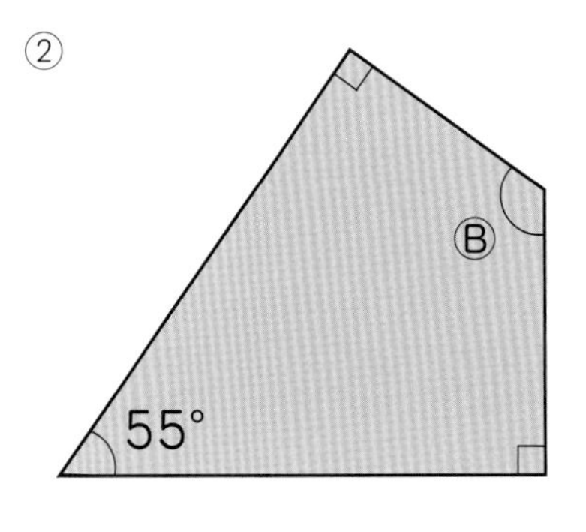

☐ We understand how to find the internal angle sum of a polygon. → pp.140 ~ 141

3 Let's fill in each ☐ with a number.

Find the internal angle sum of a heptagon.

If you draw diagonals from one vertex,

the heptagon will be divided into ☐ triangles.

Therefore, 180° × ☐ = ☐ °

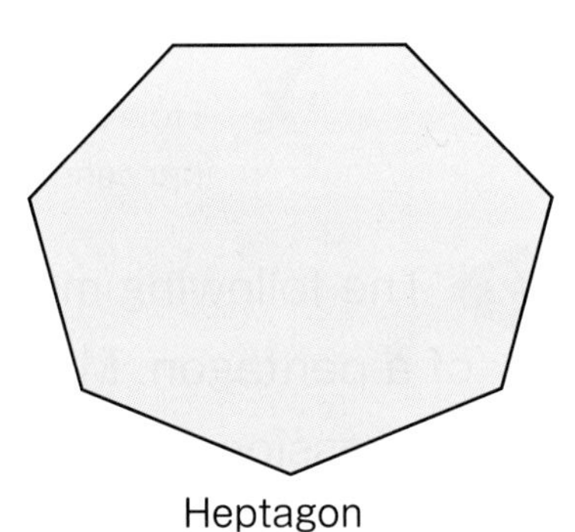

Heptagon

Supplementary Problems → p.164

Which "Way to See and Think Monsters" did you find in "9 Angles of Figures"?

I found "Rule" when I was examining the internal angle sum of a polygon.

Sara

I found other monsters, too!

Akari

Utilize Usefulness and Efficiency of Learning

Let's find out the size of the following angles Ⓐ~Ⓕ by calculations.

① 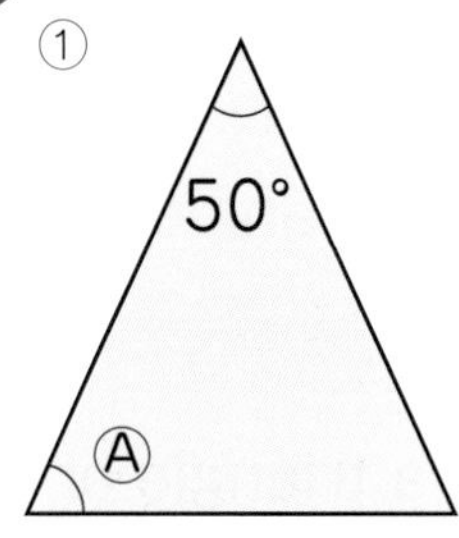 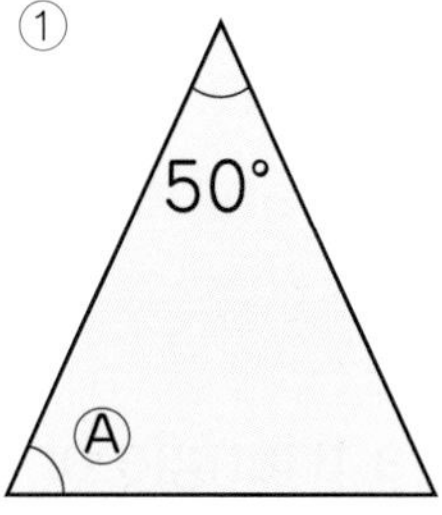

Isosceles triangle

② 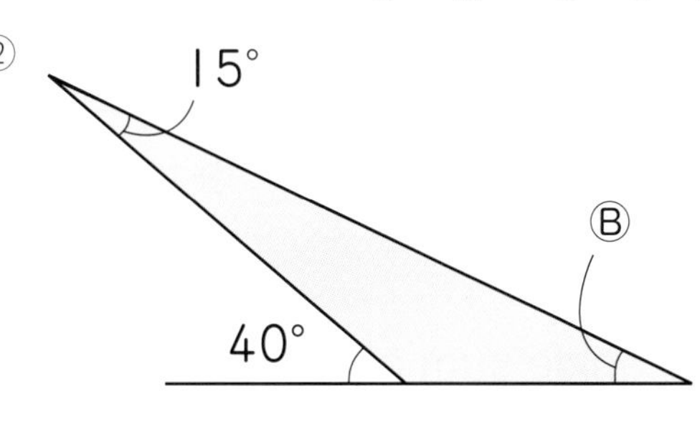

③ 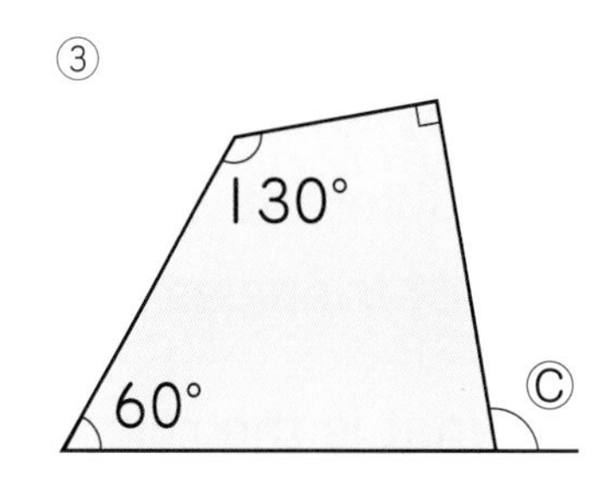

④ 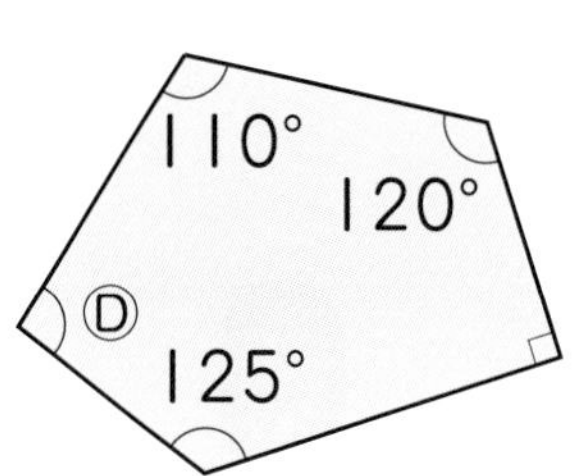

⑤ 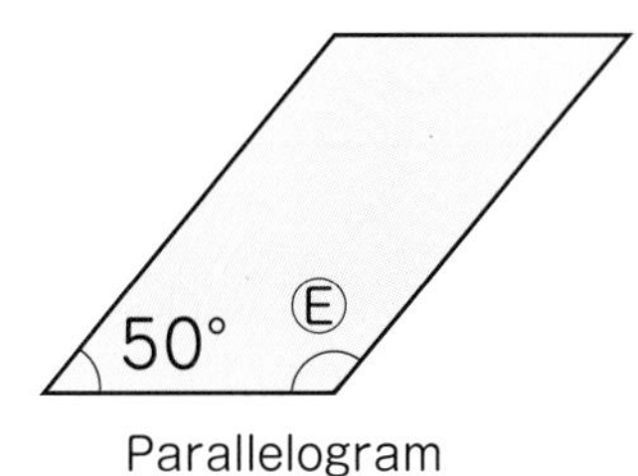

Parallelogram

⑥ Hexagon created with 6 equilateral triangles.

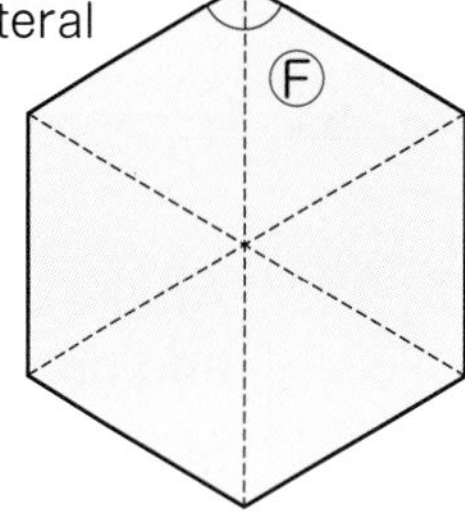

Let's find out the size of the angles Ⓐ~Ⓓ by calculations. These were made by overlapping triangle rulers.

① 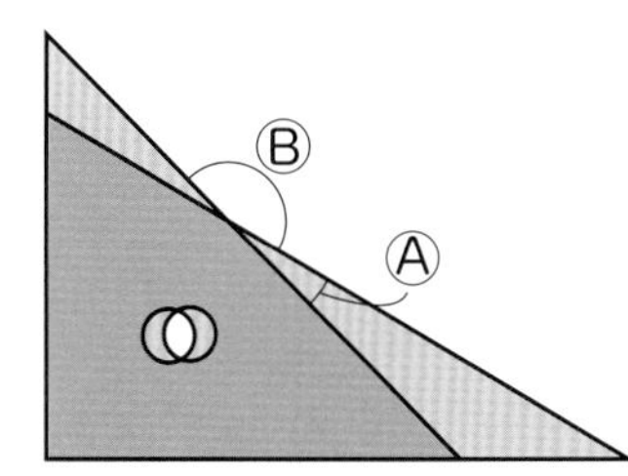

②  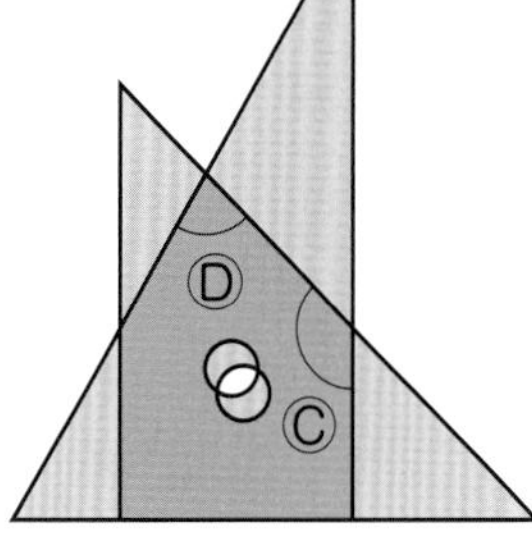

3 Let's explain how to find out the internal angle sum of an octagon using the following words.

vertex, diagonal, triangle, 180°

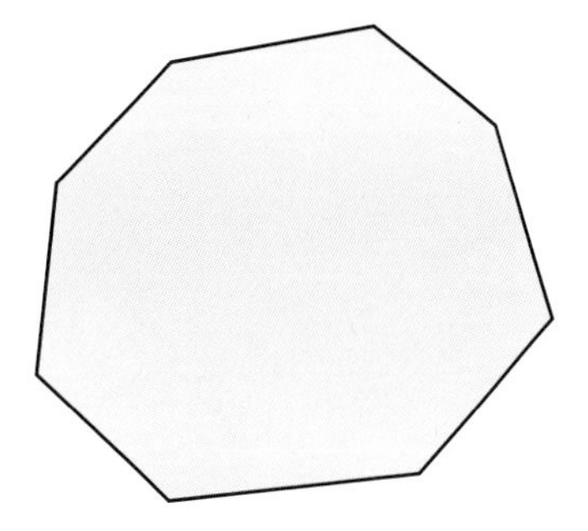

With the Way to See and Think Monsters...

Let's Reflect!

Let's reflect on which monster you used while learning " 9 Angles of Figures."

Rule

By checking the internal angle sum of triangles and polygons in different ways, we were able to find a rule for the internal angle sum of triangles and polygons.

① What is the rule for the angle sum of the three angles of a triangle or the four angles of a quadrangle?

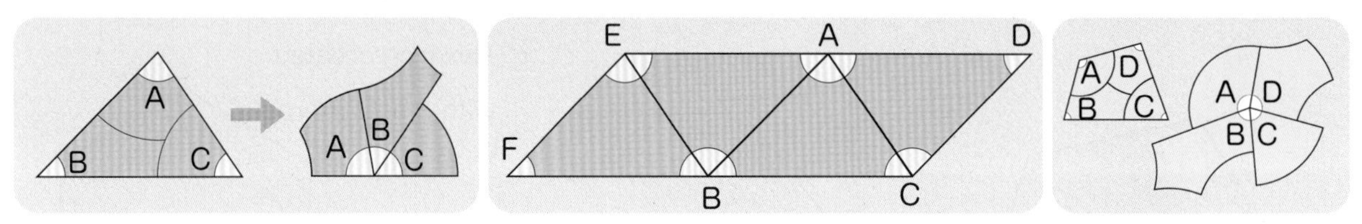

After examining the internal angle sum of various triangles and quadrilaterals, I found that the sum of the 3 angles of a triangle is ☐°, and the sum of the 4 angles of a quadrilateral is ☐°.

Sara

② What is the rule for the internal angle sum of a polygon?

Sum of the angles using the number of triangles that can be made by a diagonal drawn from one vertex

	Triangle	Quadrilateral	Pentagon	Hexagon	Heptagon	Octagon	Nonagon
Number of triangles	(1)	2	3	4	5	6	7
Angle sum	180°	360°	540°	720°	900°	1080°	1260°

Akari

In a polygon, we find the rule that the internal angle sum increases by ☐°, because for every additional angle, there is one additional triangle divided by a diagonal line drawn from one vertex.

No matter what the length of the sides are, the sum of the size of the angles is the same.

Haruto

Solve the ?

Using the fact that the internal angle sum of a triangle is 180°, we found the rule for the internal angle sum of a polygon.

Yu

Want to Connect

Are there other rules for polygons besides the internal angle sum?

Sara

Find the ?

Can we compare speeds?

▷

Who is the fastest runner in your class?

I think it's Takuya. He's a relay athlete.

Kota is fast too. He was first in the marathon.

It's hard to compare a relay with a marathon.

1

What is the fastest animal? A cheetah seems fast but...

I've heard gazelles are fast.

Birds like falcons and swallows also seem to be fast. However, it is difficult to compare the speed of running animals with animals that are flying.

2

As for vehicles, I would say the Shinkansen.

I heard the magnetic levitated trains are fast too.

Helicopters and airplanes look fast too.

3

How can we compare speeds? Can we find out in a situation that is familiar to us?

4

\ Want to know /

What should we do to compare speeds?

10 Measure per Unit Quantity (2)

Let's think about how to compare and how to represent which is faster.

How to compare speed ↓

1 The following table shows the time and distance from each house to the library. Let's explore who walks faster.

Distance and time to library

	Distance (m)	Time (min)
Kota	720	12
Haruna	660	12
Takuya	660	10

① Who is faster, Kota or Haruna?

② Who is faster, Haruna or Takuya?

Have they been walking during the whole way? I wonder if they stopped or ran at any point....
So you can think of it as the same walk from home to the library.
It does not matter how they walked, it took them a certain amount of time.
In terms of measure per unit quantity you can compare either one of the two.

You can compare who is faster if the walking time or the walking distance are matching.

When the time is the same, the one with more distance covered is faster. ▷

Kota, Haruna — Walking distance in 1 minute.

When the distances are the same, the one that takes less time is faster. ▷

Takuya, Haruna — Time needed to walk the distance.

③ Who is faster, Kota or Takuya?

Yu

The distances we took and the time it took us to get there were different.

If either were the same, we could compare them....

Akari

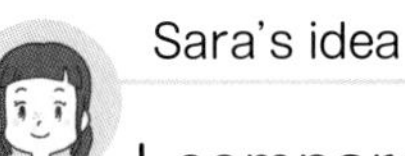

④ Let's compare the ideas of the following children.

Sara's idea

I compared the distance they walked per minute

Kota $720 \div 12 = \square$ (m)

Takuya $660 \div 10 = \square$ (m)

Therefore, $\square$ is faster.

Haruto's idea

I compared the time they walked per meter.

Kota $12 \div 720 = \square$ (min)

Takuya $10 \div 660 = \square$ (min)

Therefore, $\square$ is faster.

Yu

It is easier to understand the comparison by distance per unit of time because more quantity means faster.

Way to see and think

You've matched with either the distance walked per minute or the time taken to walk per meter.

Summary

"Who is faster?" can be compared by the time per unit of distance or by the distance per unit of time.

The speed is represented as distance per unit of time.

The math equation to find the speed : **speed = distance ÷ time**

Akari

The method for finding the speed is the same as for finding the measurement per unit quantity.

Even if you are not walking at the same speed all the time, you would assume that you were by using an average.

Haruto

⑤ Let's find out the speed of each of the 3 children.

Are speeds always compared in terms of distance per minute?

2

The Hakutaka bullet train takes 3 hours to travel 450 km between Tokyo and Kanazawa. The Hikari bullet train takes 2 hours to travel 366 km between Tokyo and Nagoya. Which bullet train is faster?

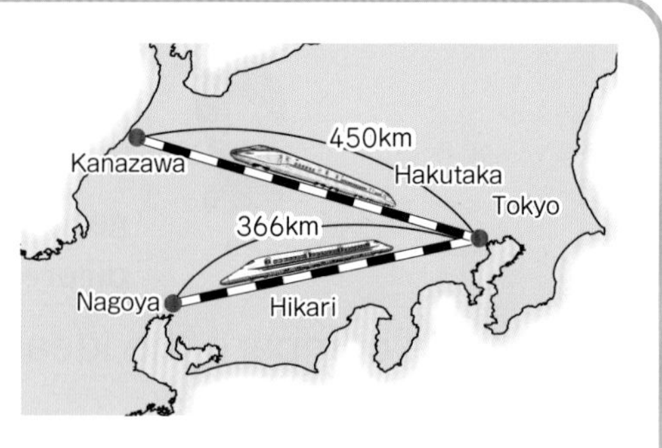

I guess we should compare the speeds.

It's a long road and a lot of time, but I guess I could convert it to minutes.

\ Want to know /

? **Purpose** How can we express the speed for long time?

① Let's find the distance that Hakutaka travels per hour.

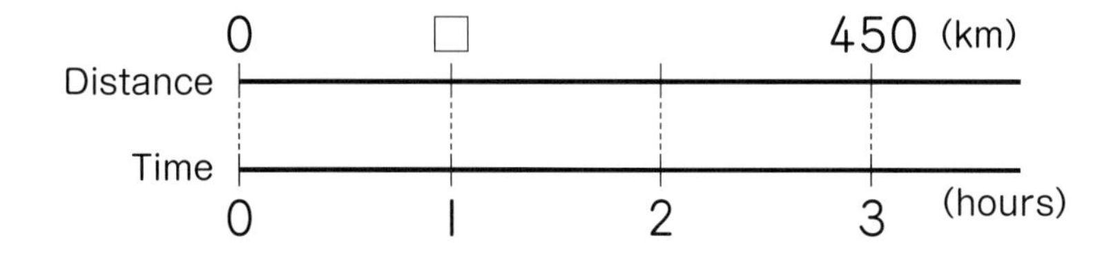

□ km	450 km
1 hour	3 hours

② Let's find the distance that Hikari travels per hour.

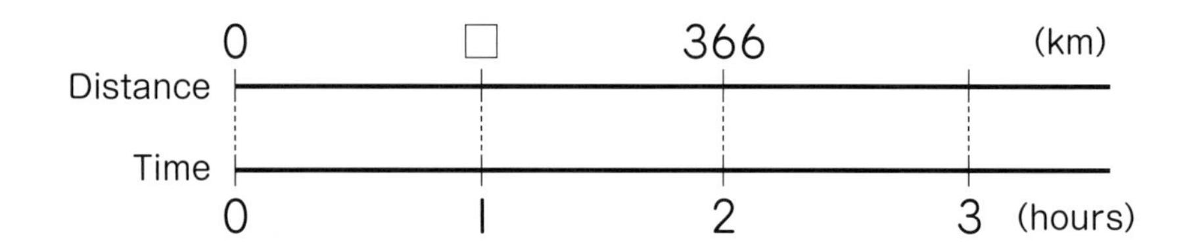

□ km	366 km
1 hour	2 hours

The speed is also a measure per unit quantity. Speed is expressed in different ways depending on the unit of time.

Speed per hour ········· Speed expressed by the distance traveled per hour.

Speed per minute ······ Speed expressed by the distance traveled per minute.

Speed per second ······ Speed expressed by the distance traveled per second.

③ What is the speed of each bullet train in kilometers per hour?

Summary Way to see and think

You can compare the speeds of long and short times by matching the unit time.

1 Who is faster, a person that runs 56 m in 8 seconds or a person that runs 60 m in 10 seconds? Let's compare the speed per second.

2 The magnetic levitated train is said to be able to travel 860 km in 2 hours. Let's find the speed per hour.

? Can the speed expressed in hours be compared to the speed expressed in seconds?

3

The news informed that the wind speed of a typhoon is 25 m per second. If a car runs at 54 km per hour. Can you say which one is faster?

The speed of the wind is also called wind speed.

I can't compare speed per hour with speed per second.

Akari

Want to think

Purpose **How can we compare speed per hour, speed per minute and speed per second?**

1 Let's explain the ideas of the following children.

Sara's idea

I compared the wind speed by changing it to speed per hour.
Since 1 minute is 60 seconds:
$25 \times 60 = 1500$, then the speed becomes 1500 m per minute.
Now, since 1 hour is 60 minutes:
$1500 \times 60 = 90000$, then the speed becomes 90000 m per hour.
Lastly, changing m to km give us 90 km per hour.

Haruto's idea

I compared the speed of the car by changing it to speed per second.
Since 1 hour is 60 minutes:
$54 \div 60 = 0.9$, then the speed becomes 0.9 km per minute, which is equivalent to 900 m per minute.
Now, since 1 minute is 60 seconds:
$900 \div 60 = 15$, then the speed is 15 m per second.

Summary

You can compare speed per hour, speed per minute, speed per second if you convert to the same quantity.

Speed per second → ×60 → Speed per minute → ×60 → Speed per hour

Speed per hour → ÷60 → Speed per minute → ÷60 → Speed per second

Per second	Per minute (60 seconds)	Per hour (60 minutes)

1 Which is the fastest among the following Ⓐ~Ⓓ?

Ⓐ a car running at 30 km/h
Ⓑ a bicycle running at 510 m/min
Ⓒ a 100-meter sprinter running at 10 m/sec
Ⓓ a camel running 36 m in 4 seconds

? Can we find the distance or the time if we know the speed?

4 **There is a car that runs at 40 km/h. How many km can it travel in 2 hours? How many km can it travel in 3 hours?**

Akari

You know how fast and how long it takes for the car.

What you are looking for is the way to proceed, right?

Yu

\ Want to think /

Can we find the distance if we know the speed and time?

① Use diagrams and tables to think about it.

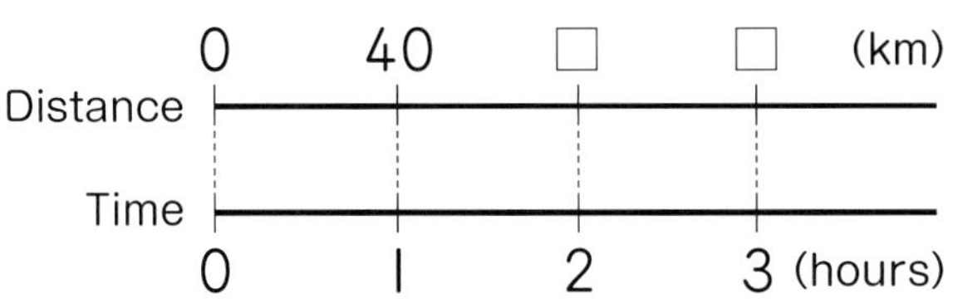

40 km	□ km
1 hour	2 hours

40 km	□ km
1 hour	3 hours

The math equation to find the distance : **distance = speed × time**

1 A cyclist travels at 400 m/min. What is the time if the cyclist travels 2400 m?

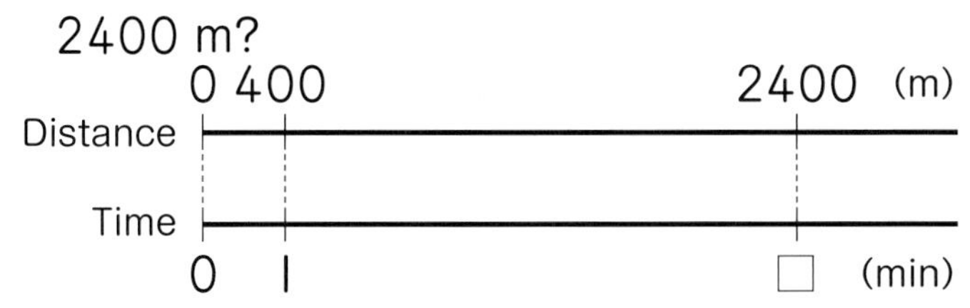

400 m	2400 m
1 min	□ min

If we know the speed and the distance, I think we can find the time.

Haruto's idea

If it takes him □ minutes, then

$400 \times \square = 2400$

$\square = 2400 \div 400$

The math equation to find the time : **time = distance ÷ speed**

If you know two among speed, distance, and time. You can use the same equation to find the missing one.

2 Let's answer the following questions about a train that travels at 30m/sec.

① What distance travels in 50 seconds?

② How long it takes in minutes to travel 5.4 km?

CAN What can you do?

☐ We understand how to find the speed. → pp.146 ~ 149

1 There is a train that travels 210 km in 3 hours. And a car that travels 160 km in 2 hours.

① What is the speed in km/h of the train?

② What is the speed in km/h of the car?

☐ We understand the relationship between speed per hour, speed per minute, and speed per second. → pp. 149 ~ 150

2 Let' s fill in the blanks in the following table and compare the speed.

	Speed per hour	Speed per minute	Speed per second
Racing car		4 km	
Helicopter ambulance	200 km		
Airplane		15 km	
Sound			340 m

☐ We understand how to find the distance. → pp.150 ~ 151

3 Let's answer the following questions.

① A typhoon is moving at 25 km/h. What distance will it move in 12 hours?

② It takes 15 minutes to travel by bus from the station to the library. If the bus travels at 42 km/h, what is the distance from the station to the library in km?

☐ We understand how to find the time. → pp.150 ~ 151

4 How many seconds will it take for a cheetah to run 180 m if the running speed is 30 m/sec?

Supplementary Problems → p.166

Which "Way to See and Think Monsters" did you find in "10 Measure per Unit Quantity (2)"?

I found "Align" when I was comparing the speed.

Akari

I found other monsters, too!

Yu

Utilize Usefulness and Efficiency of Learning

1 It takes 4 minutes for a car to pass a tunnel with a speed of 48 km/h.

① What is the speed of the car in m/min?

② What is the length of the tunnel in meters?

③ It took 2 hours and 45 minutes for this car to travel from home to the destination. What is the distance from home to the destination in km?

2 Takuto's walking speed is 60 m/min. Let's answer the following questions.

① The distance from Takuto's house to the park is 900 m. How many minutes does it take for Takuto to walk this distance?

② The distance from Takuto's house to his aunt's house is 16.2 km. How long will it take for Takuto to walk this distance?

3 The speed of sound can be called "sound speed." The sound speed at a temperature of 0 ℃ is 331 m/sec. The table below shows how the sound speed changes depending on the temperature. Now, let's consider the following problem.

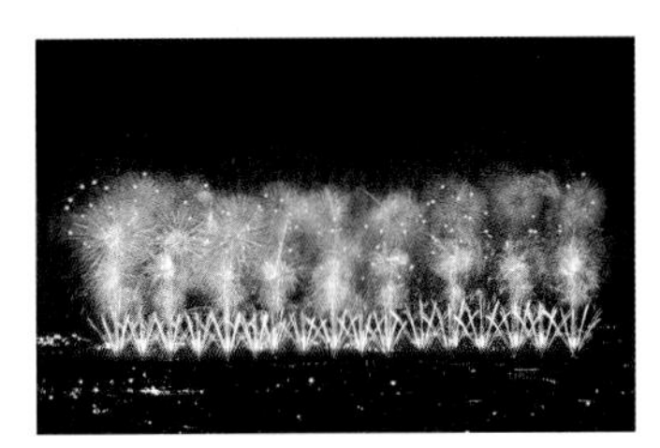

Nagaoka City, Niigata Pref.

Changes in sound speed according to temperature

Temperature (℃)	0	5	10	15	20	25	30	35
Speed per second (m/sec)	331	334	337	340	343			

① What is the speed in m/sec if the temperature increases to 5 °C?

② Daiki and Yui were watching into the sky by a window of the classroom, when a lightning flashed and after 6 seconds they heard the sound of the lightning. At that moment, the temperature was 15 °C. What was the distance to the lightning in meters?

③ Yui also saw a lightning while she was traveling. She heard the lightning striking 5 seconds after the flash. If the lightning struck 1745 m away from where she was. What was the temperature in °C at that time?

With the Way to See and Think Monsters...

Let's Reflect!

Let's reflect on which monster you used while learning " 10 Measure per Unit Quantity (2)."

Align

When finding out speeds, we could compare them in terms of time per unit distance or distance per unit time by setting the same value for either the distance or time.

① How did you find out who walks the fastest?

Distance and time to library

	Distance (m)	Time (min)
Kota	720	12
Haruna	660	12
Takuya	660	10

We can compare if either of the quantities for distance or time is the same value. If the time is the same, it is faster when the distance is ☐. If the distance is the same, it is faster if the time is ☐.

We compared in terms of the distance walked per minute.

Kota …… 720÷12=☐ (m)

Takuya … 660÷10=☐ (m)

So ☐ is faster.

We compared in terms of the time taken to walk per meter.

Kota … 12÷720=☐ (minutes)

Takuya 10÷660=☐ (minutes)

So ☐ is faster.

When neither the distance nor the time were the same, we compared them by setting the same value for the time per unit distance or the distance per unit time.

Solve the ?

Speed could be represented using the idea of measure per unit quantity.

Want to Connect

The measure per unit quantity can be obtained using the idea of how many times.
Is there any other place where we can use the idea of how many times?

More Math!

[Supplementary Problems]

[Let's deepen.]

[Answers]

Decimal Numbers and Whole Numbers

→ pp.12 ～ 19

1 Let's fill in the following □ with the appropriate answer.

① $48.3 = 10 \times \square + 1 \times \square + 0.1 \times \square$

② $0.529 = 0.1 \times \square + 0.01 \times \square + 0.001 \times \square$

③ $1 \times 7 + 0.01 \times 5 + 0.001 \times 8 = \square$

④ $0.01 \times 3 + 0.001 \times 9 = \square$

2 Let's find 10 times and 100 times of the following numbers.

① 2.46 ② 0.507 ③ 17.92 ④ 0.083

3 Let's find $\frac{1}{10}$ and $\frac{1}{100}$ of the following numbers.

① 814 ② 5.36 ③ 17.09 ④ 602.5

4 Let's fill in the following □ with the appropriate answer.

① □ times of 54.9 is 549.

② □ times of 0.286 is 28.6.

③ $\frac{1}{\square}$ of 9.3 is 0.093.

④ $\frac{1}{\square}$ of 702.1 is 70.21.

5 Let's find the answer of the following operations.

① 0.253×10 ② 4.09×100 ③ 0.086×100

④ $17.2 \div 10$ ⑤ $9.56 \div 100$ ⑥ $347.2 \div 100$

6 Let's use the digits 1, 3, 4, 7, 8 only once and the decimal point to create a decimal number with the following conditions.

① The number closest to 3 ② The number closest to 5.

Congruent Figures

→ pp.20 ～ 35

1 Which figures are congruent from the figures on the right?

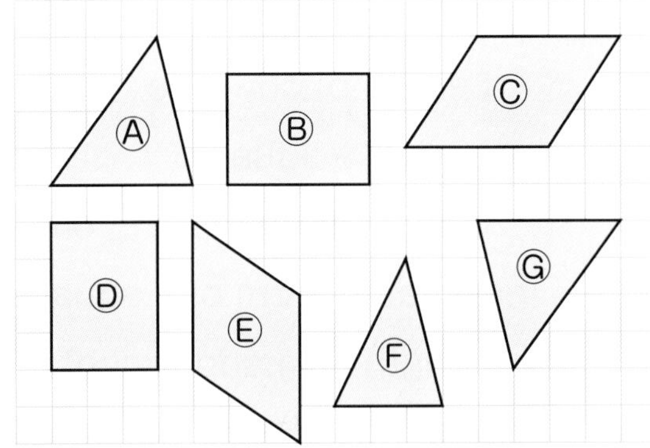

2 The following 2 quadrilaterals are congruent. Let's answer the questions below.

① Which is the corresponding vertex of the following?

ⓐ Vertex A ⓑ Vertex F

② Which is the corresponding side for the following?

ⓐ Side AB ⓑ Side FG

③ Which is the corresponding angle for the following?

ⓐ Angle D ⓑ Angle E

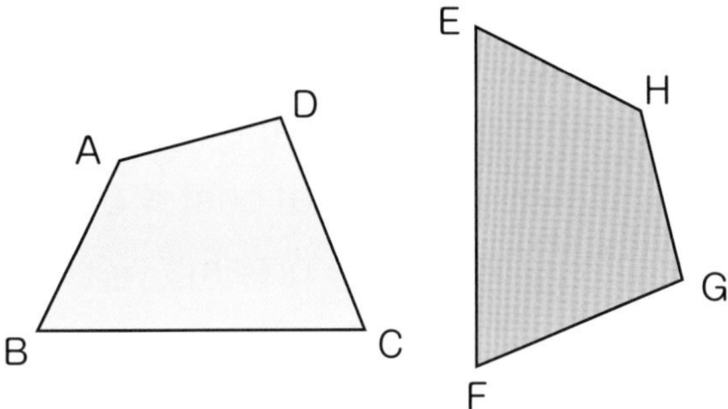

3 The 2 triangles on the right are congruent. Let's answer the questions below.

① What is the length of the following sides in cm?

ⓐ Side DE ⓑ Side DF

② What angle has equal size as the following?

ⓐ Angle B ⓑ Angle D

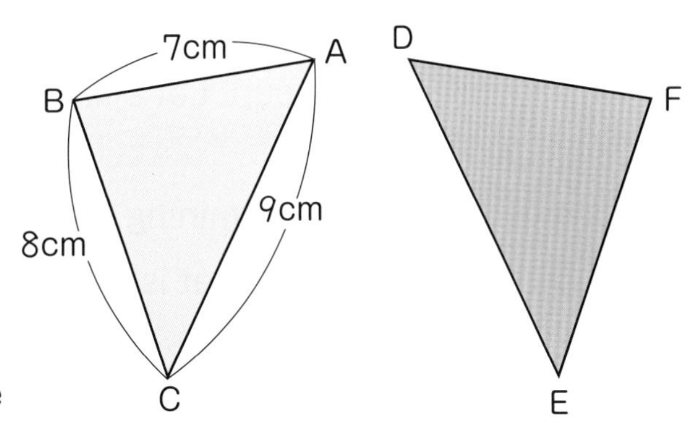

4 Let's draw a quadrilateral that is congruent to quadrilateral ABCD.

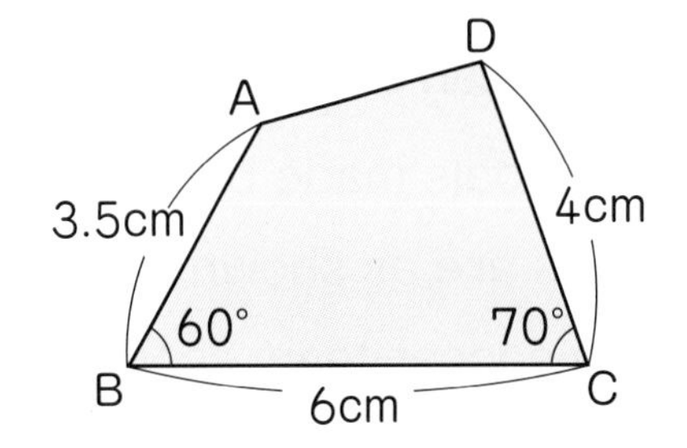

Proportion

→ pp.36 ～ 43

1 Let's choose all from ⓐ, ⓑ, and ⓒ where ○ is proportional to □.

ⓐ Length of one side □ cm and area ○ cm^2 of a square.

One side length □(cm)	1	2	3	4	5
Area ○(cm^2)	1	4	9	16	25

ⓑ Length of one side □ cm and perimeter ○ cm of a square.

One side length□(cm)	1	2	3	4	5
Perimeter ○(cm)	4	8	12	16	20

ⓒ Length □ m and weight ○ g when one meter of wire weighs 15 g.

Length □(m)	1	2	3	4	5
Weight ○(g)	15	30	45	60	75

2 Let's answer the following questions using □ number of ice cream and total cost ○ yen when one ice cream costs 80 yen.

① Let's represent the relationship between □ and ○ in a math equation.

② What is the total cost if 6 ice creams are bought?

③ How many ice creams were bought if the total cost is 960 yen?

4 Mean

→ pp.44 ～ 55

1 The table on the right shows the score for the mathematics test of 5 people in one group. Let's find the mean score.

Score in the math test

Name	Sho	Tsubasa	Karin	Yuna	Miki
Score (points)	78	87	92	75	84

2 The result after examining the weight of 6 eggs is shown on the right. What is the mean of the weight of one egg?

54 g 59 g 60 g
55 g 58 g 53 g

3 The table on the right represents the number of pages that Moka reads in 5 days. What is the mean of pages that she reads in one day?

Number of pages Moka read

Day	1st day	2nd day	3rd day	4th day	5th day
Number of pages	26	33	18	28	30

4 The goals made by a soccer team in the last 8 games are as shown on the right. How many goals did a soccer team score on average per game?

2 goals 1 goal 3 goals 0 goals
1 goal 1 goal 3 goals 1 goal

5 Multiples and Divisors

→ pp.56 ~ 73

1 Let's categorize the following numbers as even or odd numbers.

15 26 98 107 253 774

2 Let's find the first 5 multiples of the following numbers.

① 6 ② 15

3 Let's find the first 3 common multiples of the following set of numbers.

① (2, 9) ② (5, 10) ③ (6, 8)
④ (4, 10) ⑤ (2, 3, 4) ⑥ (6, 9, 12)

4 Let's find the least common multiple of the following set of numbers.

① (3, 7) ② (8, 16) ③ (12, 15) ④ (6, 10, 15)

5 Let's make a square by aligning 9 cm long and 15 cm wide rectangular paper in the same direction as shown on the right. What is the length of one side of the smallest square?

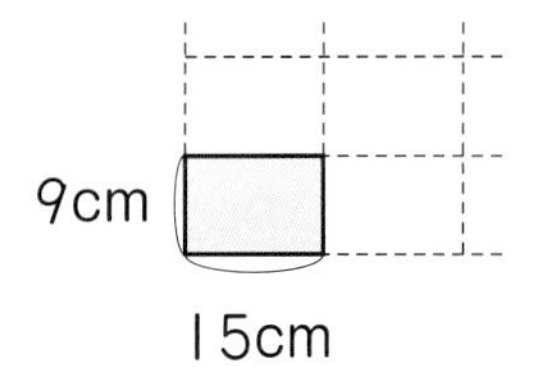

6 Let's find all the common divisors of the following set of numbers.

① (10, 20) ② (12, 18) ③ (16, 40)
④ (7, 9) ⑤ (8, 12, 16) ⑥ (15, 30, 45)

7 Let's find the greatest common divisor of the following set of numbers.

① (14, 21) ② (24, 36) ③ (27, 45) ④ (12, 24, 30)

8 There are 40 pieces of red colored paper and 32 pieces of blue colored paper. I would like to divide them among several children in such a way that each child will have the same number of pieces of each color without remainder. How many children will receive the colored paper if I divide without having remainder? Let's answer with the largest possible number of children.

Measure per Unit Quantity (1)

→ pp. 76 ~ 89

1 There are 18 children playing in the sandbox A, which has an area of 12 m^2. There are 21 children playing in the sandbox B, which has an area of 15 m^2. Which sandbox is more crowded?

2 Which train is more crowded, train ⓐ or train ⓑ?

ⓐ a train with 6 cars and 870 passengers

ⓑ a train with 9 cars and 1050 passengers

3 The following table shows the population and area of City A, City B, City C, and City D. Let's answer the following questions.

① Let's find the population density of each city as a whole number, rounded to the nearest tenth.

② Which city has the highest population density?

Population and area

	Population (people)	Area (km^2)
City A	43510	95
City B	57800	110
City C	39000	68
City D	57750	154

4 Consider a wire of 6 m long that weighs 390 g and let's answer the following questions.

① How many grams does 1 m of this wire weigh?

② What is the weight of 14 m of this wire?

③ What is the length of the wire in m when the weight of this wire is 533 g?

5 You can get 9 pencils for 720 yen or 12 pencils for 900 yen. Which pen is more expensive? Let's compare the price per unit.

6 Let's answer the following questions about a car that uses 16 L of gasoline to run 288 km.

① How many kilometers does this car run with 1 L of gasoline?

② How many kilometers does this car run with 22 L of gasoline?

③ How many liters of gasoline does this car use to run 468 km?

7 Multiplication of Decimal Numbers

→ pp.94 ～ 109

1 Let's solve the following calculations in vertical form.

① 70×3.6　② 5×1.7　③ 9×2.1　④ 17×2.4

2 What is the area in m^2 of a rectangular flower bed with a length of 4 m and a width of 2.8 m?

3 Let's solve the following calculations in vertical form.

① 3.2×2.2　② 4.7×6.6　③ 7.3×5.4　④ 1.36×1.8

⑤ 4.13×3.7　⑥ 5.1×2.64　⑦ 2.5×3.8　⑧ 1.6×4.5

⑨ 0.4×2.3　⑩ 0.3×1.9　⑪ 3.15×4.6　⑫ 5.42×3.5

4 There is a square flower bed with a length of 4.7 m. What is the area of this flower bed?

5 Let's solve the following calculations in vertical form.

① 2.6×0.4　② 0.9×0.7　③ 3.82×0.8

④ 0.65×0.2　⑤ 4.7×0.36　⑥ 0.08×0.6

6 Let's fill in the □ with the appropriate inequality sign.

① 3.8×0.9 □ 3.8　② 4.12×1.3 □ 4.12

③ 0.57×2.4 □ 0.57　④ 9.8×0.6 □ 9.8

7 There is an iron bar weighing 2.8 kg per meter. Let's answer the following questions about it.

① What is the weight in kg of this iron bar 2.5 m?

② What is the weight in kg if the iron bar has a length of 0.7 m?

8 Let's calculate by using the rules of operations. Let's also write the calculation process.

① $4.7 \times 4 \times 2.5$　② $0.6 \times 7.2 \times 5$

③ $3.6 \times 1.4 + 6.4 \times 1.4$　④ $9.1 \times 4.6 - 4.1 \times 4.6$

8 Division of Decimal Numbers → pp.110 ～ 127

1 Let's solve the following calculations in vertical form.

① 7 ÷ 1.4　② 9 ÷ 1.5　③ 63 ÷ 4.5　④ 7 ÷ 5.6

2 There is a rectangle with an area of 24 cm^2 and a length of 3.2 cm. What is the width in cm of this rectangle?

3 Let's solve the following calculations in vertical form.

① 6.72 ÷ 2.1　② 8.28 ÷ 1.8　③ 6.24 ÷ 2.4　④ 9.36 ÷ 3.6

⑤ 8.4 ÷ 2.1　⑥ 9.6 ÷ 1.6　⑦ 7.8 ÷ 1.3　⑧ 8.1 ÷ 2.7

4 There is an iron bar of 1.7 m long weighing 5.78 kg. What is the weight in kg of 1 m of this iron bar?

5 7.2 L of juice is poured into 1.8 L bottles. How many bottles were needed to pour in the juice?

6 Let's solve the following calculations in vertical form.

① 9.9 ÷ 2.2　② 6.3 ÷ 3.5　③ 51.6 ÷ 2.4　④ 55.9 ÷ 6.5

⑤ 4.42 ÷ 8.5　⑥ 3.42 ÷ 7.6　⑦ 4.93 ÷ 5.8　⑧ 6.08 ÷ 9.5

7 There is a rectangular flower bed with an area of 25.3 m^2. If the length is 4.6 m, what is the width in m?

8 There is a wire of 1.8 m with weight 22.5 g. Let's answer the following questions.

① What is the length in m of the wire per gram?

② What is the weight in g of 1 m of wire?

9 Let's solve the following calculations in vertical form.

① $28 \div 0.4$ ② $7 \div 0.8$ ③ $4.8 \div 0.6$ ④ $3.4 \div 0.5$
⑤ $1.2 \div 0.8$ ⑥ $0.3 \div 0.2$ ⑦ $0.3 \div 0.5$ ⑧ $0.1 \div 0.4$

10 Let's fill in the following □ with the appropriate equality or inequality signs.

① $1.71 \div 0.9$ □ 1.71 ② $2.72 \div 1.6$ □ 2.72
③ $0.8 \div 1$ □ 0.8 ④ $0.58 \div 0.4$ □ 0.58

11 There is a 0.8 m pipe weighing 1.16 kg. What is the weight in kg of 1 m of this pipe?

12 A piece of tape of 45 cm long is cut into 8.5 cm pieces. How many 8.5 cm pieces of tape were cut? What is the length of tape is left in cm?

13 In the following divisions, let' s find the quotient as a whole number and show the remainder.

① $22.2 \div 2.7$ ② $28.9 \div 1.9$ ③ $7.5 \div 0.7$ ④ $14.9 \div 0.6$

14 Let's solve the following divisions and round the quotient to the nearest hundredth.

① $3.1 \div 2.3$ ② $7 \div 3.8$ ③ $52.3 \div 6.8$ ④ $0.64 \div 3.4$

15 There is a car that runs 94 km with 6.5 L of gasoline. How many kilometers does this car run with 1 L of gasoline? Let's round the answer to the nearest tenth.

16 There is 5.8 L of oil weighing 5.1 kg. What is the weight of 1 L of this oil? Let's round to the nearest tenth.

17 A flower bed of 1 m^2 is watered with 2.6 L of water. Let's answer the following questions.

① What is the amount of water, in liters, that are needed to water a flower bed of 8.5 m^2?

② If 16.9 L of water is used, what is the area of this flower bed in m^2?

9 Angles of Figures

→ pp.132 ~ 144

1 Let's fill in ☐ with the appropriate answer.

① 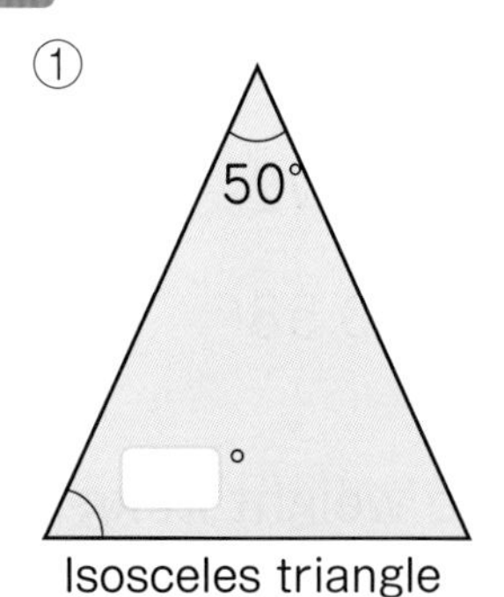

② 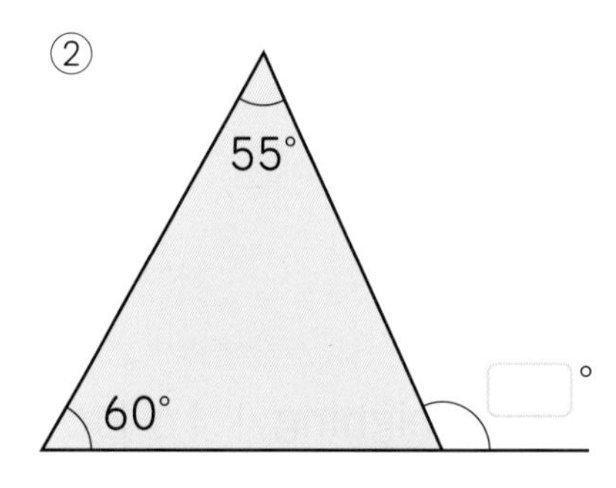

③ 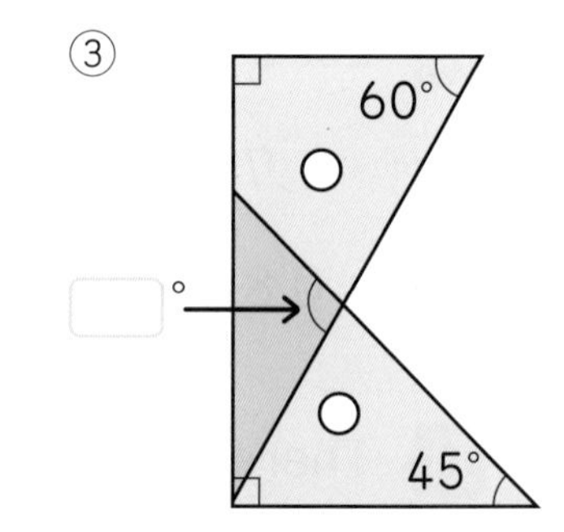

2 Let's fill in ☐ with the appropriate answer.

① The sum of the 3 angles of any triangle is ☐°.

② The sum of the 4 angles of any quadrilateral is ☐°.

3 Let's fill in ☐ with the appropriate answer.

① 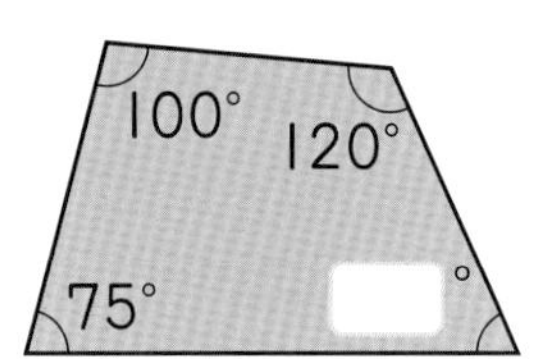

② 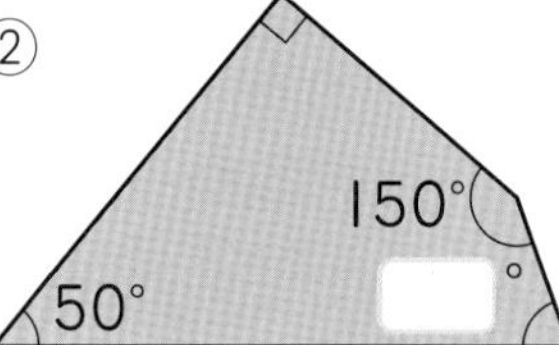

③ 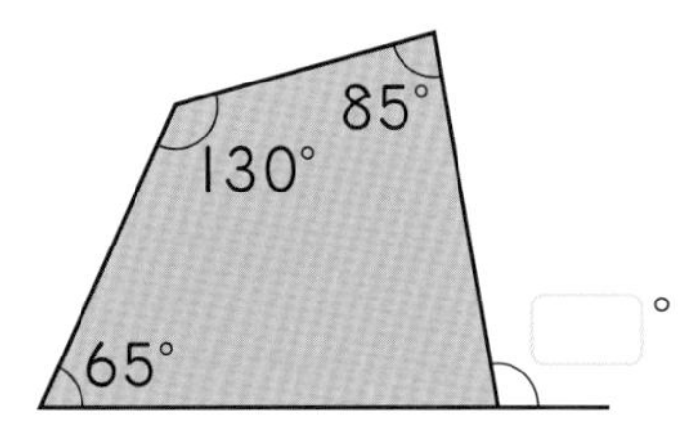

④

4 Let's fill in ☐ with the appropriate answer for the pentagon below.

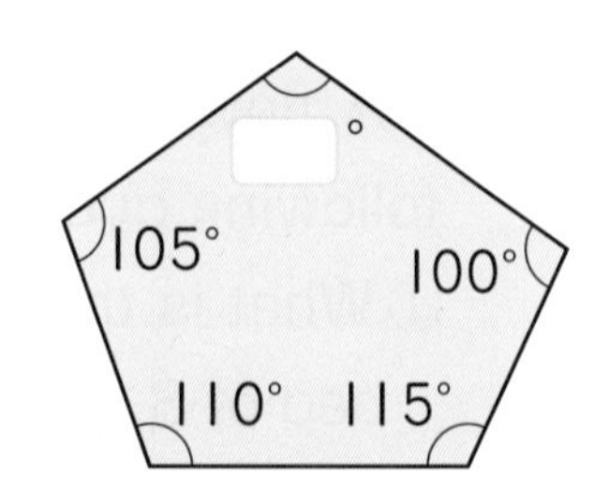

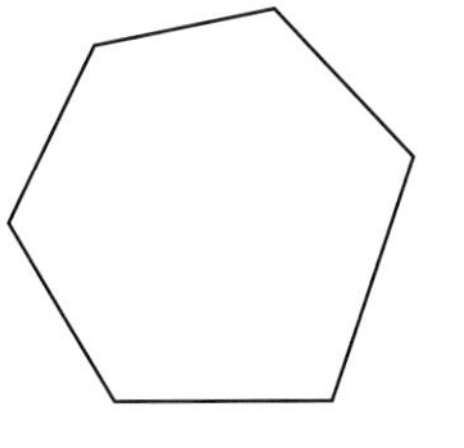

5 Let's find the sum of the 6 angles of a hexagon.

① How many diagonals can be drawn from a single vertex?

② How many triangles are made after drawing the diagonals in ①?

③ What is the sum of the 6 angles of the hexagon?

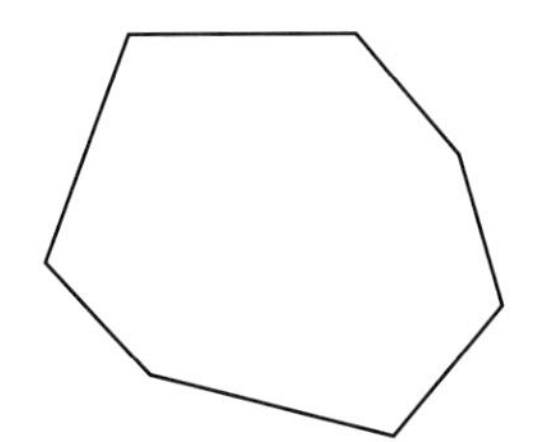

6 Let's find the sum of the 7 angles of a heptagon.

① How many diagonals can be drawn from a single vertex?

② How many triangles are made after drawing the diagonals in ①?

③ What is the sum of the 7 angles of the heptagon?

7 Let's fill in ☐ with the appropriate answer.

① The sum of the 8 angles of an octagon is, $180^\circ \times$ ☐ $=$ ☐ °.

② The sum of the 9 angles of a nonagon is, $180^\circ \times$ ☐ $=$ ☐ °.

③ The sum of the 10 angles of a decagon is, $180^\circ \times$ ☐ $=$ ☐ °.

8 Let's find the sum of the 4 angles of a quadrilateral by dividing into triangles as shown in the following figures. Let's fill in ☐ with the appropriate answer.

①

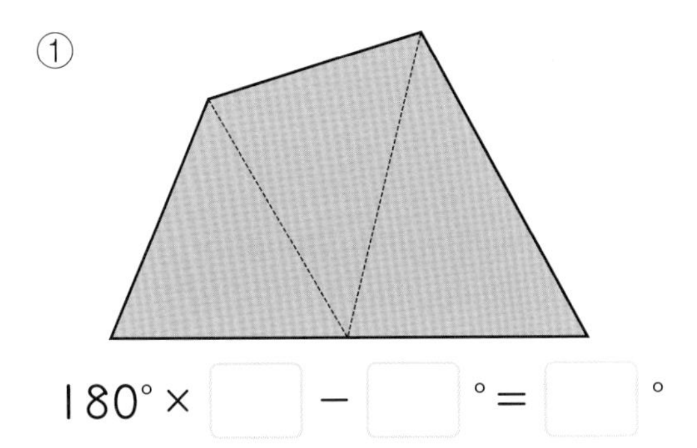

$180^\circ \times$ ☐ $-$ ☐ $^\circ =$ ☐ °

②

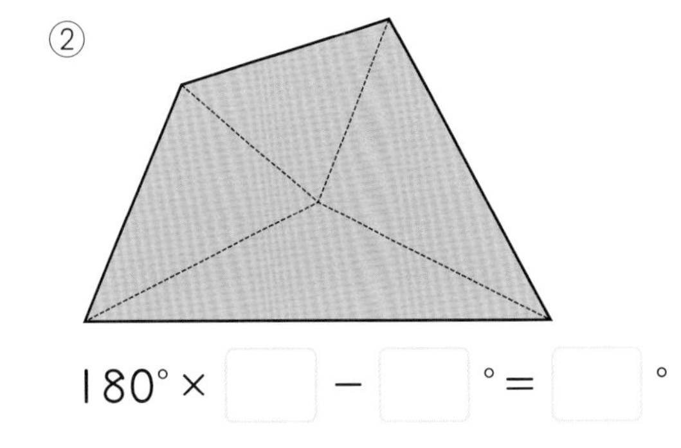

$180^\circ \times$ ☐ $-$ ☐ $^\circ =$ ☐ °

③

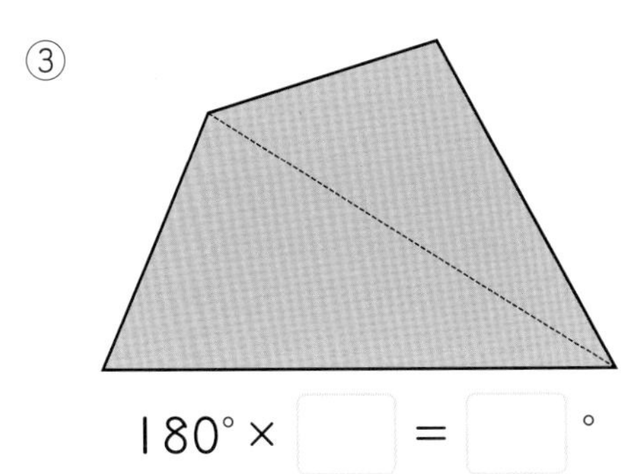

$180^\circ \times$ ☐ $=$ ☐ °

10 Measure per Unit Quantity (2)

→ pp.145 ～ 154

1 Haruma took 9 seconds to run 63 m and Ryu took 12 seconds to run 90 m. Let's answer the following questions.

① How many meters per second is the speed of Haruma and Ryu?

② Who is the fastest?

2 Let's answer the following questions.

① There is a train that travels 300 km in 4 hours. How many kilometers per hour is the speed of this train?

② There is a car that runs 21 km in 15 minutes. How many kilometers per minute is the speed of this car?

③ There is a motorcycle that runs 420 m in 30 seconds. How many meters per second is the speed of this motorcycle?

3 Let's find the following speeds.

① How many kilometers per hour is the speed of a bus traveling at 900 m per minute? And how many meters per second?

② How many kilometers per minute is the speed of a train going at 162 km per hour? And how many meters per second?

③ How many meters per minute is the speed of a person running at 5 m per second? And how many kilometers per hour?

4 There is a car going 48 km in 40 minutes. How many kilometers per hour is the speed of this car?

5 Let's find the following distances.

① A cat moves at 13 m per second. How many meters will the cat move in 6 seconds?

② A person walks at 65 m per minute. How many meters will he walk in 30 minutes?

③ A truck travels at 56 km per hour. How many kilometers will it travel in 3 hours?

④ A jet flies at 840 km per hour. How many kilometers will it travel in 2 hours?

6 Let's find the following time.

① A cheetah runs at 30 m per second. How many seconds will it take to run 195 m?

② A person runs at 280 m per minute. How many minutes will it take to run 4200 m?

③ A bicycle travels at 350 m per minute. How many minutes will it take to travel 14 km?

④ A car moves at 68 km per hour. How many hours will it take to move 204 km?

Measuring between scales

1 The temperature in the classroom was measured with a thermometer and it was between two scales as shown in the figure on the right. 9 children were asked to read the temperature individually and the table below shows the results. What is the approximate temperature of the classroom?

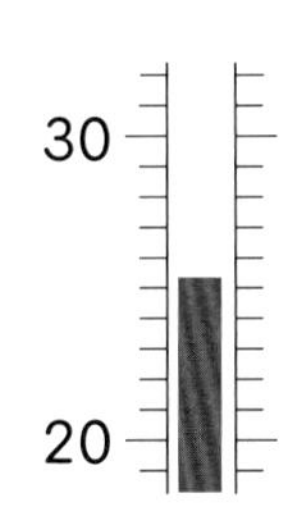

Temperature taken by the children

Name	Haruto	Arata	Akari	Mei	Yu	Rio	Daiki	Sara	Yuma
Temperature (℃)	25.2	25.1	25.3	25.2	25.2	25.1	25.4	25.2	25.1

① Let's find out the mean of the temperature taken by the 9 children.

② Let's try to discuss if the temperature found in ① is the correct temperature.

When a measurement is made with a tool and is between two scales, the mean of the readings taken by several persons may be taken as the correct result.

2 The length of one side of an equilateral triangle plate is measured with the left-hand side of the ruler at the 0 position, and the right-hand side is between two scales. 9 children read the length of each side, and the table below shows the result. Approximately how many millimeters is the length of one side of the equilateral triangle plate?

Length of one side taken by the children

Name	Haruto	Arata	Akari	Mei	Yu	Rio	Daiki	Sara	Yuma
Length (mm)	87.1	87.3	87.1	87.2	87.2	87.3	87.1	87.4	87.1

Number with two divisors

1 The number 17 only has 2 divisors: 1 and 17. From the numbers below, let's find other whole numbers with only 2 divisors: 1 and itself.

1	2	3	4	5	6	7	8	9	10
11	12	13	14	15	16	17	18	19	20
21	22	23	24	25	26	27	28	29	30
31	32	33	34	35	36	37	38	39	40

Numbers with divisors only of 1 and itself, such as 2, 3, 5, 7, 11, ..., is called a "prime number." 1 is not a prime number.

2 Follow the steps to find the prime numbers from 1 to 100.

(1) Cross the number 1.
(2) Circle number 2 and cross all the multiples of 2.
(3) Circle number 3 and cross all the multiples of 3.
⋮

The first of the remaining numbers is left and its multiples are eliminated.
Using this method, only prime numbers remain. This method is said to have been invented by the ancient Greek mathematician Eratosthenes, and is called "the sieve of Eratosthenes."

1	2	3	4	5	6
7	8	9	10	11	12
13	14	15	16	17	18
19	20	21	22	23	24
25	26	27	28	29	30
31	32	33	34	35	36
37	38	39	40	41	42
43	44	45	46	47	48
49	50	51	52	53	54
55	56	57	58	59	60
61	62	63	64	65	66
67	68	69	70	71	72
73	74	75	76	77	78
79	80	81	82	83	84
85	86	87	88	89	90
91	92	93	94	95	96
97	98	99	100		

3 Whole numbers can be expressed in the form of products of prime numbers, as $6 = 2 \times 3$. Let's express 30 and 42 in the form of products of prime numbers.

Let's think about multiples of 9.

1 Take the greatest multiple of 9 for 10 and 100 that is not greater than each of them. What is the remainder for each of them?

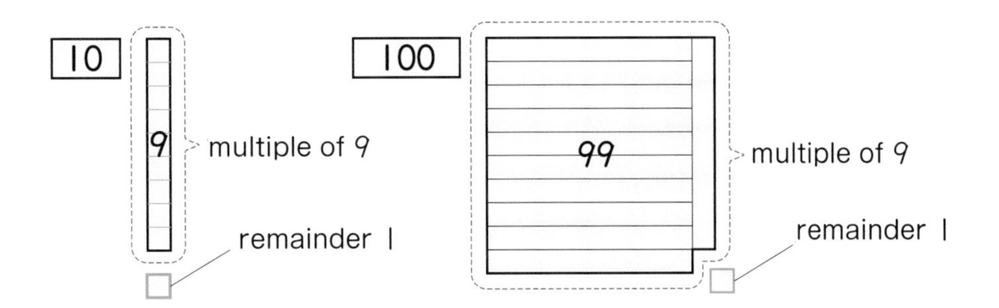

2 Let's consider whether 234 is a multiple of 9 or not.
Take the greatest multiple of 9 for 200, 30, and 4 that is not greater than each of them. What is the remainder for each of them?
What is the remainder in total? Is it a multiple of 9?

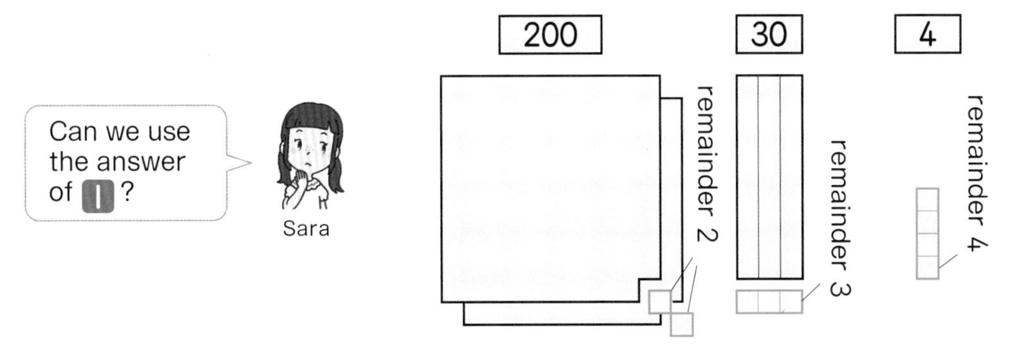

3 For other 3-digit numbers use diagrams to show that if the sum of the numbers in each place is a multiple of 9, as in 234, then the number is a multiple of 9.

If you divide 10 and 100 by 9, you get 1 more....

Haruto

Answers

[Supplementary Problems]

1 Decimal Numbers and Whole Numbers → p.156

1 ① 4,8,3 ② 5,2,9 ③ 7.058 ④ 0.039

2 ① 10 times…24.6 100 times…246
② 10 times…5.07 100 times…50.7
③ 10 times…179.2 100 times…1792
④ 10 times…0.83 100 times…8.3

3 ① $\frac{1}{10}$ …81.4 $\frac{1}{100}$ …8.14
② $\frac{1}{10}$ …0.536 $\frac{1}{100}$ …0.0536
③ $\frac{1}{10}$ …1.709 $\frac{1}{100}$ …0.1709
④ $\frac{1}{10}$ …60.25 $\frac{1}{100}$ …6.025

4 ① 10 ② 100 ③ 100 ④ 10

5 ① 2.53 ② 409 ③ 8.6
④ 1.72 ⑤ 0.0956 ⑥ 3.472

6 ① 3.1478 ② 4.8731

2 Congruent Figures → p.157

1 Ⓐ and Ⓖ, Ⓑ and Ⓓ, Ⓒ and Ⓔ

2 ① ⓐ Vertex H ⓑ Vertex C
② ⓐ Side HE ⓑ Side CD
③ ⓐ Angle G ⓑ Angle B

3 ① ⓐ 9 cm ⓑ 7 cm ② ⓐ Angle F ⓑ Angle A

4 (omitted)

3 Proportion → p.158

1 ⓑ, Ⓒ

2 ① 80 × □ = ○ ② 480 yen ③ 12 ice creams

4 Mean → p.158

1 83.2 points

2 56.5 g

3 27 pages

4 1.5 goals

5 Multiples and Divisors → p.159

1 Even numbers…26, 98, 774
Odd numbers…15, 107, 253

2 ① 6, 12, 18, 24, 30 ② 15, 30, 45, 60, 75

3 ① 18, 36, 54 ② 10, 20, 30 ③ 24, 48, 72
④ 20, 40, 60 ⑤ 12, 24, 36 ⑥ 36, 72, 108

4 ① 21 ② 16 ③ 60 ④ 30

5 45 cm

6 ① 1, 2, 5, 10 ② 1, 2, 3, 6 ③ 1, 2, 4, 8
④ 1 ⑤ 1, 2, 4 ⑥ 1, 3, 5, 15

7 ① 7 ② 12 ③ 9 ④ 6

8 8 children

6 Measure per Unit Quantity (1) → p.160

1 Sandbox ⓐ

2 Train ⓐ

3 ① City A…458 people City B…525 people
City C…574 people City D…375 people
② City C

4 ① 65 g ② 910 g ③ 8.2 m

5 9 pencils for 720 yen

6 ① 18 km ② 396 km ③ 26 L

7 Division of Decimal Numbers → p.161

1 ① 252 ② 8.5 ③ 18.9 ④ 40.8

2 11.2 m^2

3 ① 7.04 ② 31.02 ③ 39.42 ④ 2.448
⑤ 15.281 ⑥ 13.464 ⑦ 9.5 ⑧ 7.2
⑨ 0.92 ⑩ 0.57 ⑪ 14.49 ⑫ 18.97

4 22.09 m^2

5 ① 1.04 ② 0.63 ③ 3.056
④ 0.13 ⑤ 1.692 ⑥ 0.048

6 ① < ② > ③ > ④ <

7 ① 7 kg ② 1.96 kg

8 ① 47 ② 21.6 ③ 14 ④ 23

8 Division of Decimal Numbers → pp.162 ~ 163

1 ① 5 ② 6 ③ 14 ④ 1.25
2 7.5 cm
3 ① 3.2 ② 4.6 ③ 2.6 ④ 2.6
⑤ 4 ⑥ 6 ⑦ 6 ⑧ 3
4 3.4 kg
5 4 bottles
6 ① 4.5 ② 1.8 ③ 21.5 ④ 8.6
⑤ 0.52 ⑥ 0.45 ⑦ 0.85 ⑧ 0.64
7 5.5 m
8 ① 0.08 m ② 12.5 g
9 ① 70 ② 8.75 ③ 8 ④ 6.8
⑤ 1.5 ⑥ 1.5 ⑦ 0.6 ⑧ 0.25
10 ① $>$ ② $<$ ③ $=$ ④ $>$
11 1.45 kg
12 5 pieces, 2.5 remained
13 ① 8 remainder 0.6 ② 15 remainder 0.4
③ 10 remainder 0.5 ④ 24 remainder 0.5
14 ① 1.35 ② 1.84 ③ 7.69 ④ 0.19
15 Around 14.5 km
16 Around 0.9 kg
17 ① 22.1 L ② 6.5 m^2

9 Angles of Figures → pp.164 ~ 165

1 ① 65 ② 115 ③ 105
2 ① 180 ② 360
3 ① 65 ② 70 ③ 100 ④ 120
4 110
5 ① 3 ② 4 ③ 720°
6 ① 4 ② 5 ③ 900°
7 ① 6, 1080 ② 7, 1260 ③ 8, 1440
8 ① 3, 180, 360 ② 4, 360, 360 ③ 2, 360

10 Measure per Unit Quantity (2) → p.166

1 ① Haruma…7 m per second
Ryu…7.5 m per second
② Ryu
2 ① 75 km per hour ② 1.4 km per minute
③ 14 m per second
3 ① 54 km per hour, 15 m per second
② 2.7 km per minute, 45 m per second
③ 300 m per minute, 18 km per hour
4 72 km per hour
5 ① 78 m ② 1950 m
③ 168 km ④ 1680 km
6 ① 6.5 seconds ② 15 minutes
③ 40 minutes ④ 3 hours

[Let's deepen.]

Measuring between scales → p.167

1 ① Approximately 25.2 ℃ ② (omitted)
2 Approximately 87.2 mm

ADVANCED Number with two divisors → p.168

1 2, 3, 5, 7, 11, 13, 17, 19, 23, 29, 31, 37
2 (omitted)
3 $30 = 2 \times 3 \times 5$
$42 = 2 \times 3 \times 7$

Let's think about multiples of 9. → p. 169

1 10…1 100…1
2 200…2 30…3 4…4
There are 9 remainders in total, and is the multiple of 9.
3 (omitted)

words
which we learned in this textbook

Congruent Figures

→ To be used in page 21.
Please cut these out for use.

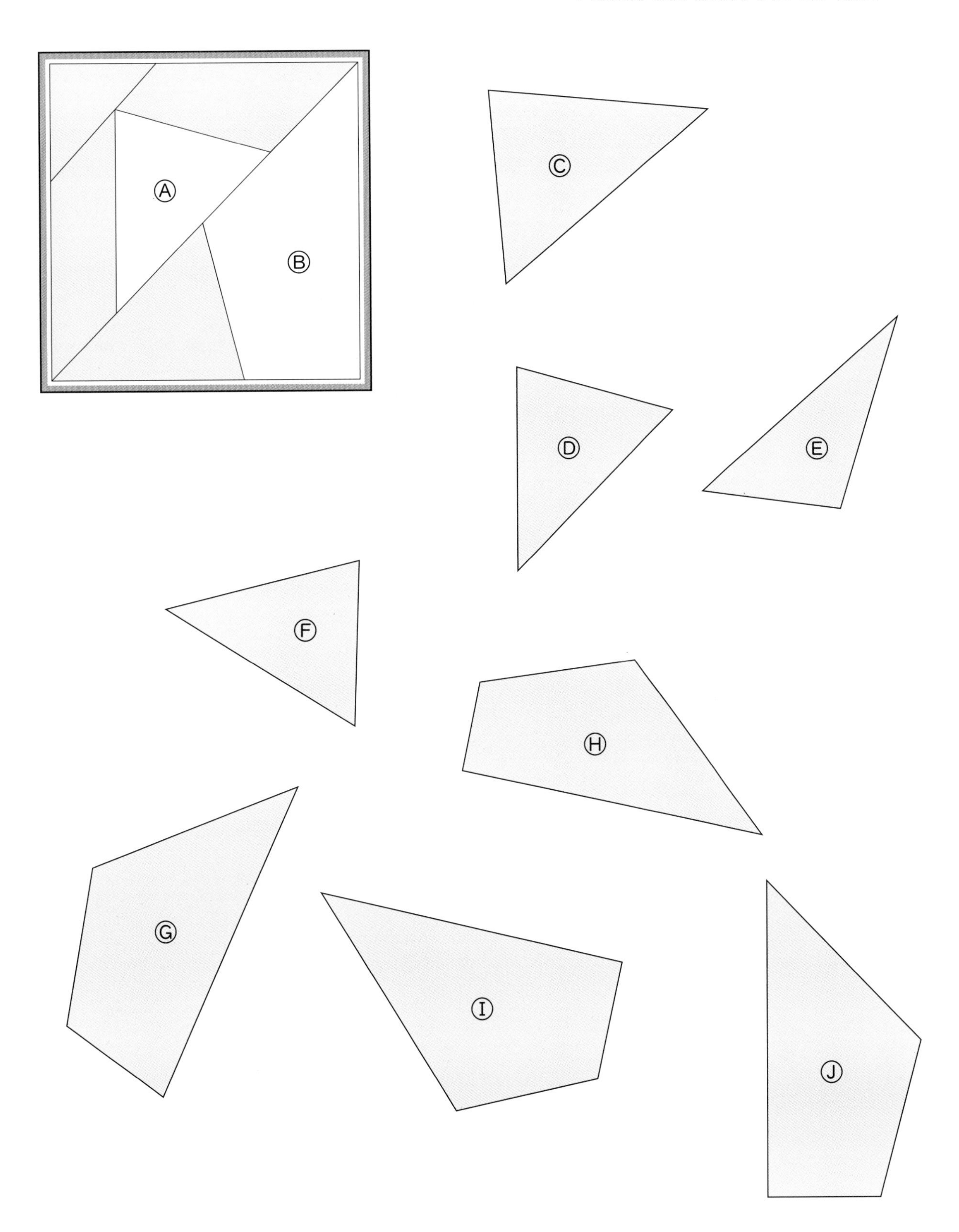

Angles of Figures

→ To be used in pages 137 and 138.
Please cut these out for use.

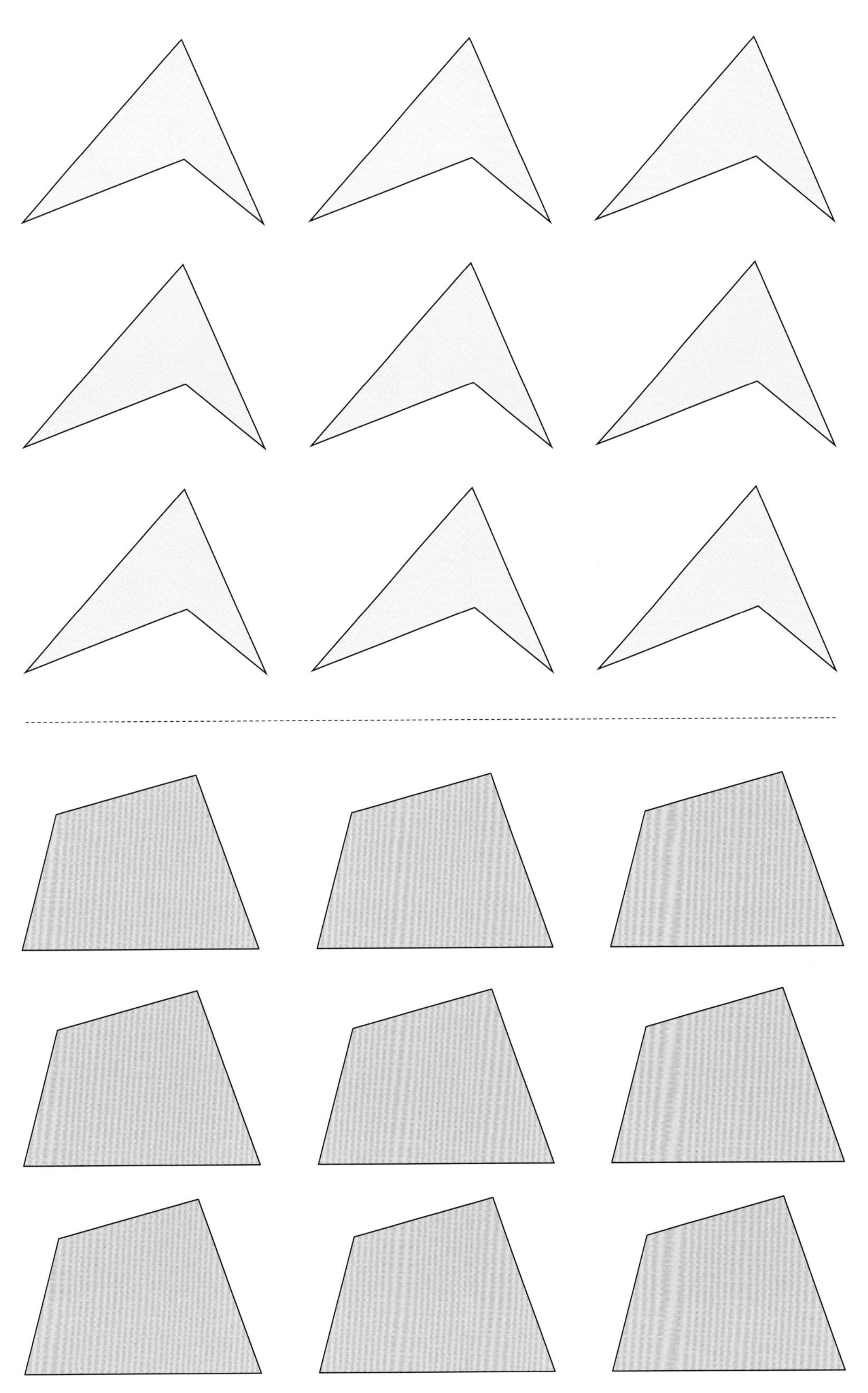

Memo

Memo

Memo

Editors of Original Japanese Edition

[Head of Editors]

Shin Hitotsumatsu (Kyoto University), Yoshio Okada (Hiroshima University)

[Supervising Editors]

Toshiyuki Akai (Hyogo University), Toshikazu Ikeda (Yokohama National University), Shunji Kurosawa (The former Rikkyo University), Hiroshi Tanaka (The former Tsukuba Univ. Elementary School), Kosho Masaki (The former Kokugakuin Tochigi Junior College), Yasushi Yanase (Tamagawa University)

[Editors]

Shoji Aoyama (Tsukuba Univ. Elementary School), Kengo Ishihama (Showa Gakuin Elementary School), Hiroshi Imazaki (Hiroshima Bunkyo University), Atsumi Ueda (Hiroshima University), Tetsuro Uemura (Kagoshima University), Yoshihiro Echigo (Tokyo Gakugei Univ. Setagaya Elementary School), Hisao Oikawa (Yamato University), Hironori Osawa (Yamagata University Graduate School), Tomoyoshi Owada (Shizuoka University), Nobuhiro Ozaki (Seikei Elementary School), Masahiko Ozaki (Kansai Univ. Elementary School), Kentaro Ono (Musashino University), Hiroshi Kazama (Fukui University), Michihiro Kawasaki (Oita University), Miho Kawasaki (Shizuoka University), Yoshiko Kambe (Tokai University), Yukio Kinoshita (Kwansei Gakuin Elementary School), Tomoko Kimura (Tamon Elementary School, Setagaya City), Satoshi Kusaka (Naruto University of Education), Kensuke Kubota (Naruo Higashi Elementary School, Nishinomiya City), Itsushi Kuramitsu (The former University of the Ryukyus), Maiko Kochi (Kounan Elementary School, Toshima City), Chihiro Kozuki (Daiyon. Elementary School, Hino City), Goto Manabu (Hakuoh University), Michihiro Goto (Tokyo Gakugei Univ. Oizumi Elementary School), Hidenori Kobayashi (Hiroshima Univ. Shinonome Elementary School), Akira Saito (Shibata Gakuen University Graduate School), Masahiko Sakamoto (The former Tokoha University Graduate School), Junichi Sato (Kunitachigakuen Elementary School), Hisatsugu Shimizu (Keio Yochisha Elementary School), Ryo Shoda (Seikei University), Masaaki Sugihara (University of the Sacred Heart, Tokyo), Jun Suzuki (Gakushuin Primary School), Shigeki Takazawa (Shiga University), Chitoshi Takeo (Nanzan Primary School), Hidemi Tanaka (Tsukuba Univ. Elementary School), Toshiyuki Nakata (Tsukuba Univ. Elementary School), Hirokazu Nagashima (Daisan. Elementary School, Kokubunji City), Minako Nagata (Futaba Primary School), Kiyoto Nagama (Hiyagon Elementary School, Okinawa City), Satoshi Natsusaka (Tsukuba Univ. Elementary School), Izumi Nishitani (Gunma University), Kazuhiko Nunokawa (Joetsu University of Education), Shunichi Nomura (Waseda University), Mantaro Higuchi (Kori Nevers Gakuin Elementary School), Satoshi Hirakawa (Showa Gakuin Elementary School), Kenta Maeda (Keio Yokohama Elementary School), Hiroyuki Masukawa (University of the Sacred Heart, Tokyo), Shoichiro Machida (Saitama University), Keiko Matsui (Hasuike Elementary School, Harima Town), Katsunori Matsuoka (Naragakuen University), Yasunari Matsuoka (Matsushima Elementary School, Naha City), Yoichi Matsuzawa (Joetsu University of Education), Satoshi Matsumura (Fuji Women's University), Takatoshi Matsumura (Tokoha University), Kentaro Maruyama (Yokohama National Univ. Kamakura Elementary School), Kazuhiko Miyagi (Homei Elementary School Affiliated with J.W.U), Aki Murata (University of California, Berkeley), Takafumi Morimoto (Tsukuba Univ. Elementary School), Yoshihiko Moriya (The former Kunitachigakuen Elementary School), Junichi Yamamoto (Oi Elementary School, Oi Town), Yoshikazu Yamamoto (Showa Gakuin Elementary School), Shinya Wada (Kagoshima University), Keiko Watanabe (Shiga University)

[Reviser]

Katsuto Enomoto (The former Harara Elementary School, Kagoshima City)

[Reviser of Special Needs Education and Universal Design]

Yoshihiro Tanaka (Teikyo Heisei University)

[Cover]

Photo : Tatsuya Tanaka (MINIATURE LIFE)

Design : Ai Aso (ADDIX)

[Text]

Design : Ai Aso, Takayuki Ikebe, Hanako Morisako, Miho Kikuma, Sawako Takahashi, Katsuya Imamura, Satoko Okutsu, Kazumi Sakaguchi, Risa Sakemoto (ADDIX), Ayaka Ikebe

[Illustrations]

Lico, Nanami Sumimoto, Ryoko Totsuka, Kinue Naganawa, Yoshie Hayashi, B-rise

[Photo・Observation data]

Ritmo, Pixta, Pacific Press Service, Shiiba Village Tourism Association, Yamaha Corporation, Honda Motor Co., Ltd., Chobitto 19+, JR Central, Niigata Prefecture Tourism Association